London Mathematic Society Lecture Note Series: 57

Techniques of Geometric Topology

ROGER A. FENN

Lecturer, Mathematics Division, University of Sussex

CAMBRIDGE UNIVERSITY PRESS

Cambridge

London New York New Rochelle

Melbourne Sydney

Published by the Press Syndicate of the University of Cambridge
The Pitt Building, Trumpington Street, Cambridge CB2 1RP
32 East 57th Street, New York, NY 10022, USA
296 Beaconsfield Parade, Middle Park, Melbourne 3206, Australia

First published 1983

Printed in Great Britain at the University Press, Cambridge

Library of Congress catalogue card number: 81-18189

British Library Cataloguing in Publication Data

Fenn, Roger A.
Techniques of geometric topology. - (London Mathematical Society
lecture note series, ISSN 0076-0552; 57)
1. Topology
I. Title
514 QA611

ISBN 0 521 28472 4

To Marion, Tom, Georgina and James.

LONDON MATHEMATICAL SOCIETY LECTURE NOTE SERIES

Managing Editor: Professor I.M. James,
Mathematical Institute, 24-29 St Giles, Oxford

1. General cohomology theory and K-theory, P.HILTON
4. Algebraic topology, J.F.ADAMS
5. Commutative algebra, J.T.KNIGHT
8. Integration and harmonic analysis on compact groups, R.E.EDWARDS
9. Elliptic functions and elliptic curves, P.DU VAL
10. Numerical ranges II, F.F.BONSALL & J.DUNCAN
11. New developments in topology, G.SEGAL (ed.)
12. Symposium on complex analysis, Canterbury, 1973, J.CLUNIE & W.K.HAYMAN (eds.)
13. Combinatorics: Proceedings of the British Combinatorial Conference 1973, T.P.McDONOUGH & V.C.MAVRON (eds.)
15. An introduction to topological groups, P.J.HIGGINS
16. Topics in finite groups, T.M.GAGEN
17. Differential germs and catastrophes, Th.BROCKER & L.LANDER
18. A geometric approach to homology theory, S.BUONCRISTIANO, C.P. ROURKE & B.J.SANDERSON
20. Sheaf theory, B.R.TENNISON
21. Automatic continuity of linear operators, A.M.SINCLAIR
23. Parallelisms of complete des[illegible] J.CAMERON
24. The topology of Stiefel man[illegible] I.M.JAMES
25. Lie groups and [illegible] PRICE
26. Transformatio[illegible]gs of the conference in the University of Newcastle-[illegible] 1976, C.KOSNIOWSKI
27. Skew field con[illegible]
28. Brownian motion, [illegible] bounded mean oscillations, K.E.PETERSEN
29. Pontryagin duali[illegible] tructure of locally compact Abelian groups, S.A.MORRI[illegible]

Interaction model[illegible] GGS

Continuous crosse[illegible] ts a[illegible] III von Neumann algebras, A.VAN DAELE

Uniform algebras [illegible], T.W.GAMELIN

Permutation groups [illegible] l structures, N.L.BIGGS & A.T.WHITE

Representation theo[illegible], M.F. ATIYAH et al.

Trace ideals and t[illegible], B.SIMON

Homological group th[illegible] ALL (ed.)

Partially ordered ri[illegible] lgebraic geometry, G.W.BRUMFIEL

Surveys in combinato[illegible] BAS (ed.)

Affine sets and affine [illegible] D.G.NORTHCOTT

Introduction to Hp spaces, [illegible] KOOSIS

Theory and applications of [illegible] bifurcation, B.D.HASSARD, N.D.KAZARINOFF & Y-H.WAN

Topics in the theory of gr[illegible] presentations, D.L.JOHNSON

Graphs, codes and designs, [illegible].CAMERON & J.H.VAN LINT

Z/2-homotopy theory, M.C.CRABB

Recursion theory: its generalisations and applications, F.R.DRAKE & S.S.WAINER (eds.)

p-adic analysis: a short course on recent work, N.KOBLITZ

Coding the Universe, A.BELLER, R.JENSEN & P.WELCH

Low-[illegible]

49. Finite geometries and designs, P.CAMERON, J.W.P.HIRSCHFELD & D.R.HUGHES (eds.)
50. Commutator calculus and groups of homotopy classes, H.J.BAUES
51. Synthetic differential geometry, A.KOCK
52. Combinatorics, H.N.V.TEMPERLEY (ed.)
53. Singularity theory, V.I.ARNOLD
54. Markov processes and related problems of analysis, E.B.DYNKIN
55. Ordered permutation groups, A.M.W.GLASS
56. Journées arithmétiques 1980, J.V.ARMITAGE (ed.)
57. Techniques of geometric topology, R.A.FENN
58. Singularities of smooth functions and maps, J.MARTINET
59. Applicable differential geometry, M.CRAMPIN & F.A.E.PIRANI
60. Integrable systems, S.P.NOVIKOV et al.
61. The core model, A.DODD
62. Economics for mathematicians, J.W.S.CASSELS
63. Continuous semigroups in Banach algebras, A.M.SINCLAIR
64. Basic concepts of enriched category theory, G.M.KELLY
65. Several complex variables and complex manifolds I, M.J.FIELD
66. Several complex variables and complex manifolds II, M.J.FIELD
67. Classification problems in ergodic theory, W.PARRY & S.TUNCEL
68. Complex algebraic surfaces, A.BEAUVILLE
69. Representation theory, I.M.GELFAND et al.
70. Stochastic differential equations on manifolds, K.D.ELWORTHY
71. Groups - St Andrews 1981, C.M.CAMPBELL & E.F.ROBERTSON (eds.)
72. Commutative algebra: Durham 1981, R.Y.SHARP (ed.)
73. Riemann surfaces: a view towards several complex variables, A.T.HUCKLEBERRY
74. Symmetric designs: an algebraic approach, E.S.LANDER
75. New geometric splittings of classical knots (algebraic knots), L.SIEBENMANN & F.BONAHON
76. Linear differential operators, H.O.CORDES
77. Isolated singular points on complete intersections, E.J.N.LOOIJENGA
78. A primer on Riemann surfaces, A.F.BEARDON
79. Probability, statistics and analysis, J.F.C.KINGMAN & G.E.H.REUTER (eds.)
80. Introduction to the representation theory of compact and locally compact groups, A.ROBERT
81. Skew fields, P.K.DRAXL
82. Surveys in combinatorics: Invited papers for the ninth British Combinatorial Conference 1983, E.K.LLOYD (ed.)
83. Homogeneous structures on Riemannian manifolds, F.TRICERRI & L.VANHECKE
84. Finite group algebras and their modules, P.LANDROCK
85. Solitons, P.G.DRAZIN
86. Topological topics, I.M.JAMES (ed.)
87. Surveys in set theory, A.R.D.MATHIAS (ed.)

CONTENTS

PREFACE

The purpose of this book is to explain the basic techniques of algebraic topology with illustrations and examples from low dimensional topology. I hope that this book will prove useful to all students with a smattering of algebraic topology who wish to learn more, but I also hope that even the experts will find something in here to interest them.

In Chapter 1 I have tried to revise the homology theory and notation needed for the later chapters. Not many details of proofs are given as they may be found in any competent book on homology theory. In this chapter, as in the rest of the book, I have tried to explain some of the geometric ideas behind homology theory, in particular the idea of a mock bundle and the relationship between cup products and the dual notion of intersection.

In Chapter 2 I have tried to do the same thing for homotopy theory as Chapter 1 does for homology theory.

Chapter 3 is concerned with covering spaces. In particular the universal covering of surfaces is treated, as this is usually not dealt with in the standard texts.

Chapter 4 is concerned with the homology poperties of covering spaces. I have also included a section on the lower central series as this will be useful later.

In Chapter 5 I have attempted to describe the theory of knots and links and how they illustrate the theory developed in the previous chapters.

Finally, Chapter 6 is an introduction to the esoteric theory of Massey products. This is partly the fruits of collaboration with Denis Sjerve of U.B.C.

It would be very difficult and time-consuming to delineate all the colleagues who have helped me in my understanding of this subject and the writing of this book. So I hope they will forgive me if they are not mentioned explicitly here. However, I should acknowledge the enormous debt of gratitude which I owe to the following:

John Reeve for acting as my Ph.D. tutor and getting me started on the right road. Colin Rourke, who with infinite impatience has explained to me on many occasions great wodges of mathematics in his own inimitable style. Martin Dunwoody, for teaching me about groups and group presentations, and for proof reading several of the chapters. Richard Lewis for his help with sections of Chapter Four. All the mathematicians at U.B.C. for their kindness and hospitality during my sabbatical visit when this book was started. In particular, I would like to thank James Whittaker who invited me in the first place, Mark Goresky for revealing the secrets of the double banana, Denis Sjerve for his helpful explanations of many aspects of algebraic topology and Dale Rolfsen, whose books on knots and links were an inspiration for the whole project.

Last but not least, I must express my thanks and admiration to Jill Foster for producing such a beautiful typescript.

1. AN INTRODUCTION TO HOMOLOGY

'It (homology theory) is the oldest and most extensively developed portion of algebraic topology, and may be regarded as the main body of the subject.'

Eilenberg and Steenrod *Foundations of Algebraic Topology*

1.1 Cell complexes

A cell is modelled on a ball in Euclidean space and a cell complex is built up from cells subject to certain conditions which we will consider later. The resulting combinatorial structure is a device for calculation. Rather like a basis of a vector space, it has no special merit in itself.

Definition 1.1.1 Cells and cell complexes.

Euclidean space of dimension n is denoted by R^n. A typical point of R^n is an n-triple $x = (x_1, \ldots, x_n)$ of real numbers x_i, $i = 1, 2, \ldots, n$. The Pythagorean metric is given by the formula

$$\| x - y \| = \left\{ \sum_{i=1}^{n} (x_i - y_i)^2 \right\}^{\frac{1}{2}} .$$

The n-ball of radius 1 in R^n is defined by

$$B^n = \{ x \in R^n \mid \| x \| \leq 1 \}.$$

Note that B^1 is twice as long as the unit interval $I = [0, 1]$.

The boundary of the n-ball is the (n-1)-sphere defined by

$$S^{n-1} = \{ x \in R^n \mid \| x \| = 1 \}.$$

So S^0 is a pair of points, S^1 is a circle and S^2 is a sphere. The circle S^1 can also be considered as the set of unit complex numbers $e^{i\theta}$.

If X is a topological space and $f : S^{i-1} \to X$ is a continuous map then the identification space $X \cup_f B^i$ is obtained by identifying the points $f(b)$ and b of the disjoint union of X with B^i, where b lies in S^{i-1}.

We say that $X \cup_f B^i$ is obtained from X by **attaching a cell of dimension i**. The image of B^i is called an **i-cell**. The resulting space is written $X \cup \sigma^i$ where σ^i is the image of B^i. The map f is called the **attaching map** of the cell σ^i and the natural identification map $\phi : B^i \to X \cup \sigma^i$ is called the **characteristic map** of the cell σ^i. Note that ϕ is 1-1 on the interior of B^i, $B^i - S^{i-1}$, but is not necessarily so on the boundary S^{i-1}. Let $\dot{\sigma}^i = f(s^{i-1})$ and $int\ \sigma^i = \sigma^i - \dot{\sigma}^i$.

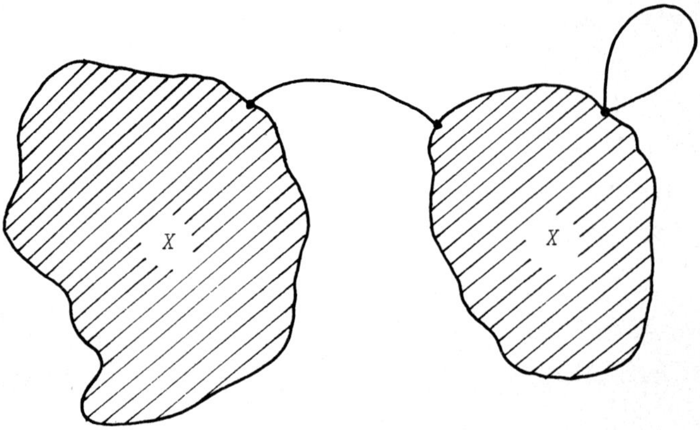

Figure 1. A space X with two 1-cells attached.

A **cell complex** K or just **complex** for short is a filtered space $K = \bigsqcup_{n=0}^{\infty} K^n$, $K^0 \subset K^1 \subset K^2 \subset \ldots$.

Each K^n is called the **n-skeleton** of K and is obtained by attaching n-cells to K^{n-1}. So K^n consists of all cells of dimension n or less and

$$int\ \sigma_i \cap int\ \sigma_j = \phi \quad if\ \ \sigma_i \neq \sigma_j .$$

Note that since *n-cells* are attached to K^{n-1} they can be attached in any order.

Two more conditions will be imposed on cell complexes in this book [1]. These are:

(1) K^o the *o-skeleton* is a discrete space;

(2) K has only countably many cells and each cell meets only finitely many other cells.

Sometimes it will be useful to emphasise the **underlying space** $|K| = \bigcup_{\sigma \in K} \sigma$ which has a natural topological structure as an identification space. However, for the most part K will stand for both the space $|K|$ and the combinatorial structure of cells.

The space $|K|$ is said to be **decomposed by** K or **decomposed into** K. The **dimension**, $dim\ K$, of K is the maximum dimension of any one of its cells and may be infinite.

Note that each n-*cell* as the image of B^n has a built-in orientation.

Examples. Cell complexes.

1. The n-sphere S^n can be decomposed into a cell complex with one *o-cell* and one *n-cell*. So

$$S^n = \sigma^o \cup \sigma^n$$

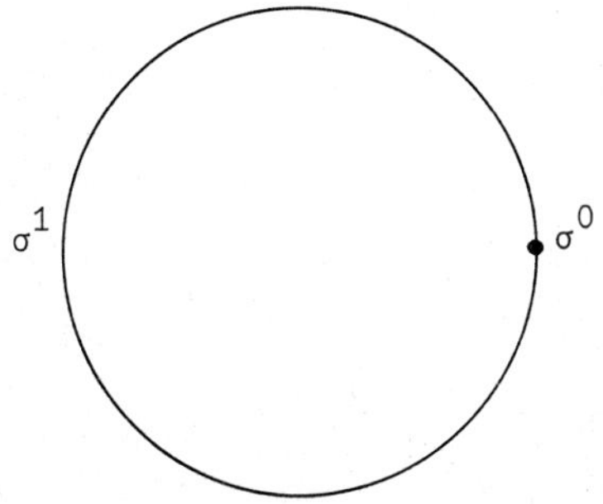

Figure 2. S^1 as a cell complex

2. If K, L are cell complexes then $K \times L$ is the **product complex** with skeleta defined by:

$$(K \times L)^n = \bigcup \left\{ \sigma^i \times \sigma^j \mid \sigma^i \in K, \ \sigma^j \in L, \ i + j \leq n \right\}.$$

So for example the **2-torus** $S^1 \times S^1$ can be decomposed into one *o-cell* $\sigma^0 \times \sigma^0$, two *1-cells* $\sigma^0 \times \sigma^1$, $\sigma^1 \times \sigma^0$ and one *2-cell* $\sigma^1 \times \sigma^1$

3. If K, L are disjoint cell complexes each with one specified *o-cell* then $K \vee L$ their **wedge** or **one point union** is the result obtained by identifying these *o-cells*. For example, if K is the decomposition of the torus considered above then K^1 is $S^1 \vee S^1$, a figure eight.

4. A collection of cells in a complex K which themselves form a complex is called a **subcomplex** of K. If $L \subset K$ is a subcomplex then (K, L) is called a **cell complex pair** or **pair** for short. For example, (K^n, K^m) is a pair if $n \geq m$.

5. If K is a simplicial complex then K may be considered as a cell complex if each of its simplexes is oriented. A convenient way of doing this is to linearly order the vertices of each simplex. Usually this is done so that the ordering is respected by the faces of each simplex.

The following useful lemma says that the attaching map of any cell can be changed by a homotopy without changing the homotopy type of the complex. The proof is a standard application of the homotopy extension property for complexes, see for example Whitehead, 1949.

Lemma 1.1.2 *(The cell homotopy lemma)* [2]

Let the attaching map $f : S^{i-1} \to K^i$ *of an i-cell be changed by a homotopy.*

Then K *is deformed into another complex which has the same homotopy type.*

Corollary 1.1.3 *Any finite connected cell complex* K *has the homotopy type of a cell complex with just one 0-cell.*

Proof. Suppose that K has two *0-cells* σ^0 and ρ^0 joined by a *1-cell* σ^1. Consider the homotopy of the attaching map of σ^1 which shrinks σ^1 to σ^0. The resulting complex has the form $L \vee I$ where I is the *1-complex* with two *0-cells* σ^0, ρ^0 and one *1-cell* ρ^1. But $L \vee I$ has the homotopy type of L which has one fewer *0-cell* than K, [3].

1.2 2-complexes and group presentations.

A cell complex of dimension 2 with just one *0-cell* is of particular interest. Let e be the *0-cell* to which *1-cells* $a_1, a_2, \ldots$ are attached to form a rosette of loops homeomorphic to $S^1 \vee S^1 \vee \ldots$, see figure 3.

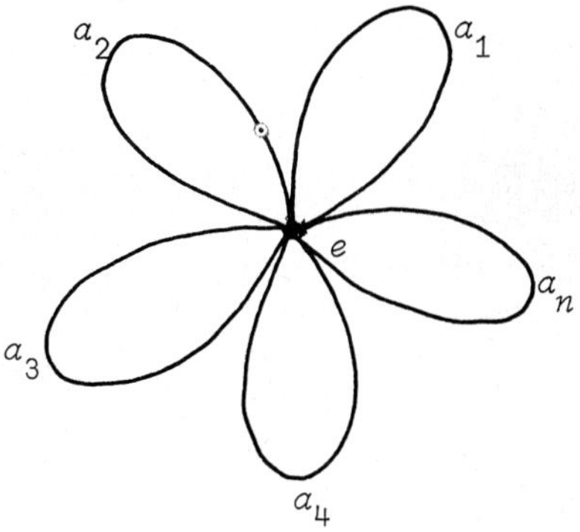

Figure 3. *1-skeleton of a cell complex with just one 0-cell.*

Let $r_1, r_2, \ldots$ be the *2-cells* and let $f : S^1 \to K$ be the attaching map of one of them. Then by the cell homotopy lemma this map may be altered by homotopy without changing the homotopy type of K. In fact we can suppose that after a homotopy the inverse image $f^{-1} e$ is equal to the nth roots of unity, so

$$f^{-1}e = \left\{1, \omega^1, \omega^2, \ldots, \omega^{n-1}\right\}, \qquad \omega = e^{2\pi i/n} .$$

Then if (ω^i, ω^{i+1}) is the open-ended arc of S^1 between ω^i and ω^{i+1}, $i = 1, 2, \ldots, n$, by a further homotopy we can make $f \mid (\omega^i, \omega^{i+1})$ a 1-1 map.

Let $a_{\alpha(i)}$ be the *1-cell* traversed by $f(\omega^i, \omega^{i+1})$ in this manner. Then associated with the *2-cell* is a word (also denoted by r) of the form

$$r = a_{\alpha(1)}^{\varepsilon(1)} \, a_{\alpha(2)}^{\varepsilon(2)} \cdots a_{\alpha(n)}^{\varepsilon(n)} ,$$

where the index $\varepsilon(i)$ is $+1$ for a positive (anticlockwise) traversal and -1 for a negative traversal. Up to a cyclic re-ordering of their elements the words r define K and this definition is unique up to homotopy equivalence.

Definition 1.2.1 Group presentations.

If A is a set let A^{-1} be the collection of formal inverses. So A and A^{-1} are disjoint sets in bijective correspondence by the rule $a \leftrightarrow a^{-1}$.

A group presentation is a pair of sets $\{A \mid R\}$ such that associated with each r in R is a word

$$\omega(r) = x_1 x_2 \ldots x_k, \qquad x_i \in A \cup A^{-1} .$$

Strictly speaking the relators r are distinct from the relations $\omega(r)$ although they will very often be confused. So, for example, $\{a \mid r\}$ where $\omega(r) = a^n$, $\{a \mid a^n\}$, $\{a \mid a^n = 1\}$ and even $\{a \mid r = a^n\}$ are all different ways of writing the same presentation of the cyclic group of order n. For more details see Karras, Magnus and Solitar.

So associated with each 2-complex with one *0-cell* is a group presentation and conversely. In what follows no distinction will be made between the two notions.

Examples. Group presentations.

1. $\{a \mid r, s\}$ $\omega(r) = \omega(s) = a$, is the *2-sphere* S^2 and is also a presentation of the trivial group.
2. $\{\phi \mid 1\}$ is also the *2-sphere*.
3. $\{a \mid 1\}$ is $S^1 \vee S^2$.
4. Write $[a, b] = aba^{-1}b^{-1}$ then $\{a, b \mid [a, b]\}$ is the *2-torus* $S^1 \times S^1$ and presents the torus group $\mathbb{Z} \oplus \mathbb{Z}$.
5. Any closed compact surface can be decomposed as either

$$T_\gamma = \{a_1, \ldots, a_\gamma, b_1, \ldots, b_\gamma \mid [a_1, b_1] \ldots [a_\gamma, b_\gamma]\} \quad \text{or}$$
$$P_\gamma = \{a_1, a_2, \ldots, a_\gamma \mid a_1^2 a_2^2 \ldots a_\gamma^2\}.$$

Here T_γ is the orientable surface of genus γ, $\gamma = 0, 1, 2, \ldots,$ and P_γ is the non-orientable surface of genus γ, $\gamma = 1, 2, \ldots,$ [4].

1.3 Homology of cell complexes.

As was mentioned in the introduction, it will be assumed that the reader has some acquaintance with homology theory and the seven axioms of Eilenberg and Steenrod, [5]. These axioms might be called the definition of homology. It is shown in Eilenberg and Steenrod that a homology theory exists and is unique at least for cell complex pairs. The calculation of these homology groups is, in principle at least, straightforward and is affected by the use of cellular chains. Before introducing these, some notation and a definition is given.

The homology groups of the pair (K, L) with coefficients in G are denoted by $H_n(K, L; G)$, $n = 0, 1, \ldots$ [6]. If the G is omitted it will be assumed that the coefficient group is $\mathbb{Z}$. If $f : (K, L) \to (K_1, L_1)$ is a continuous map then the induced homomorphism is written f_*. The symbol for the map will be dropped in any situation where the map is obvious, e.g. inclusion.

Definition 1.3.1 Incidence number.

Let σ^i, σ^{i+1} be cells in K of adjacent dimension. By reordering it can be assumed that σ^i is the last *i-cell* attached to K^{i-1} to form K^i and σ^{i+1} is the first *(i+1)-cell* attached to K^i. Then $K^i = A \cup \sigma^i$ where A is K^{i-1} with all the previous *i-cells* attached. Let $f : S^i \to A \cup \sigma^i$ be the attaching map of σ^{i+1}. Let $[\sigma^i ; \sigma^{i+1}]$ denote the degree of the composition

$$S^i \xrightarrow{f} A \cup \sigma^i \longrightarrow A \cup \sigma^i / A \cong S^i . \quad [7]$$

The integer $[\sigma^i ; \sigma^{i+1}]$ is called the **incidence number** of σ^i with σ^{i+1}.

Example.

Let $K = \{A \mid R\}$ be a *2-complex* with one *0-cell* e and let $\varepsilon_a(r)$ denote the total exponent of a in the word $\omega(r)$, [8]. Then $[e ; a] = 0$ for all a in A and $[a ; r] = \varepsilon_a(r)$.

Definition 1.3.2 Cellular chains.

Let $C_n(K)$ be the free abelian group with basis the *n-cells* of K. Let $C_n(K, L) = C_n(K) / C_n(L)$. [9].

The elements of $C_n(K)$ are called **cellular n-chains** of K. The elements of $C_n(K, L)$ are called **cellular n-chains** of K *mod* L. The boundary map $\partial : C_n(K, L) \to C_{n-1}(K, L)$ is defined on the basis of *n-cells* by

$$\partial \sigma^n = \sum [\sigma^{n-1} ; \sigma^n] \, \sigma^{n-1} ,$$

the sum being taken over all *(n-1)-cells* σ^{n-1} of K. [10].

Theorem 1.3.3

1. *The composition* $C_{n+1}(K, L) \xrightarrow{\partial} C_n(K, L) \xrightarrow{\partial} C_{n-1}(K, L)$ *is zero.*

2. $H_n(K, L) \simeq Ker\{C_n \xrightarrow{\partial} C_{n-1}\} / Im\{C_{n+1} \xrightarrow{\partial} C_n\}$, [11].

Proof. See Cooke and Finney.

The elements of $\mathrm{Ker}\{C_n \to C_{n-1}\}$ are called **n-cycles.** The elements of $\mathrm{Im}\{C_{n+1} \to C_n\}$ are called **n-boundaries.**

If H_n has a basis $\{[z_1], \ldots, [z_n]\}$ where $z_1, \ldots, z_n$ are cycles then $\{z_1, \ldots, z_n\}$ is called a **cycle basis** for H_n.

The above theorem refers to homology with $\mathbb{Z}$ coefficients, although a similar theorem holds for homology with G coefficients. The chain complex being given by:

$$\ldots \longrightarrow C_n(K, L) \otimes G \xrightarrow{\partial \otimes 1} C_{n-1}(K, L) \otimes G \longrightarrow \ldots$$

The importance of 1.3.3 lies in the fact that however $|K|$ is decomposed by K the group $H_p(K)$ always equals $H_p(|K|)$.

Examples.

1. For any complex K, $H_0 K$ is the free abelian group of rank ρ, where ρ is the number of components of K.

2. Let $K = \{A \mid R\}$ be a *2-complex* with one *0-cell*. Then $H_1 K$ has the presentation

$$\sum_{a \in A} \varepsilon_a(r)\, a = 0, \qquad r \in R,$$

and $H_2(K)$ is the free abelian group with rank

$$\#(R) - \mathit{rank}\,(\varepsilon_a(r)).$$

For example, if $T = \{a, b \mid [a, b]\}$, the torus, then $H_0(T) \approx H_2(T) \approx \mathbb{Z}$ and $H_1(T) \approx \mathbb{Z} \oplus \mathbb{Z}$. A cycle basis for $H_1(T)$ is given by the cycles a and b. If $P = \{a \mid a^2\}$, the projective plane, then $H_0(P) \approx \mathbb{Z}$, $H_1(P) \approx \mathbb{Z}_2$, and $H_2(P) \approx 0$. However, $H_1(P; \mathbb{Z}_2) \approx H_2(P; \mathbb{Z}_2) \approx \mathbb{Z}_2$.

If the *2-cells* of K can be oriented to form a cycle then this cycle is often denoted by K and its resulting homology class by $[K]$.

1.3.4 Geometric chains.

The original idea of homology due to Poincaré, Veblen and others was to make an algebra based on manifolds. It would indeed be pleasant if all homology classes could be represented by manifolds. However, theory has shown this to be false, although it is possible to represent homology classes by something which is almost a manifold.

Definition 1.3.5 Geometric chains.

An n-circuit (or manifold with codimension 2 singularities) is a complex K with the following properties:

1. Every point of K either lies in an $(n-1)$-*cell* or in the interior of an n-*cell*.

2. Every n-*cell* and $(n-1)$-*cell* is homeomorphic to a ball.

3. An $(n-1)$-*cell* is the face of one or two n-*cells*. The closure of the $(n-1)$-*cells* satisfying the former condition is a subcomplex ∂K, the boundary of K.

4. The n-*cells* of K are oriented to form an integer chain c such that $|\partial c| = |\partial K|$.

5. The boundary ∂K is an $(n-1)$-*circuit*.

So the complex K is an n-*manifold* except for a subcomplex (the singularities) of dimension $\leq n-2$.

A geometric n-chain in a space X consists of an n-*circuit* K and a map $f : K \to X$.

If $f(\partial K) \subset A$ then $f : K, \partial K \to X, A$ represents some class in $H_n(X, A)$. Before considering the converse theorem it is necessary to recall the definitions of singular homology.

Definition 1.3.6 Singular Homology.

Let $\nu_0 = (0,0,0,\ldots, 0)$ be the origin in R^n and let $\nu_1 = (1,0,\ldots, 0)$, $\nu_2 = (0,1,\ldots, 0),\ldots,$ $\nu_n = (0,0,\ldots, 1)$ be the unit points. The standard n-simplex Δ_n is the convex hull of the $n+1$ points $\nu_0, \nu_1,\ldots, \nu_n$.

Consider the simplicial maps

$$\partial_i : \Delta_{n-1} \longrightarrow \Delta_n \qquad i = 0,1,\ldots, n,$$

given by the following vertex assignments:

$$\partial_i(\nu_j) = \begin{cases} \nu_j & \text{if } j < i \\ \nu_{j+1} & \text{if } j \geq i. \end{cases}$$

So ∂_i maps Δ_{n-1} onto the *(n-1)-face of* Δ_n not containing v_i and preserves the order of the vertices.

A **singular n-simplex** in X is a continuous map

$$f : \Delta_n \longrightarrow X .$$

The singular *(n-1)-simplex*

$$f_{(i)} = f\partial_i : \Delta_{n-1} \longrightarrow X$$

is called the ith face of f.

Let $C_n(X)$ denote the free abelian group generated by the singular *n-simplexes* in X. The **boundary** homomorphism

$$\partial : C_n \longrightarrow C_{n-1}$$

is defined on singular simplexes by the rule

$$\partial f = \sum_{i=0}^{n} (-1)^i f_{(i)} .$$

The groups C_n together with the boundary homomorphisms form a chain complex and the resulting homology theory is called **singular homology.** It satisfies the axioms and is therefore the unique theory for cell complexes. More details can be found in Eilenberg and Steenrod.

The singular theory is totally useless as a calculating tool, but it is a very convenient theoretical device, as the following theorem shows:

Theorem 1.3.7 *Any singular homology class in* $H_n(X, A)$ *can be represented by a geometric n-chain* $f : K, \partial K \longrightarrow X, A$.

The idea behind the proof is to use a singular chain representing the class to define the geometric chain.

Let the class be represented by the singular chain

$$c = \sum_{i=1}^{r} \lambda_i f_i ,$$

where the λ_i are non-zero integers and the f_i are singular *n-simplexes* in X.

Let Y be the disjoint union of $|\lambda_1| + |\lambda_2| + \ldots + |\lambda_r|$ n-*simplexes*. Partition Y into k blocks $\{B_i\}_{i=1}^{k}$ where B_i contains $|\lambda_i|$ n-*simplexes*, $i = 1, \ldots, r$. For each n-*simplex* σ in Y let $\alpha_\sigma : \sigma \longrightarrow \Delta_n$ be a simplicial isomorphism matching σ with the standard n-*simplex*. Consider the collection of all pairs (σ, s), where σ is an n-*simplex* of Y and s is a face of σ. Call two pairs (σ, s) and (τ, t) **identifiable** if the following conditions hold:

1. if $\sigma \in B_i$, $\tau \in B_j$ then $i \neq j$.
2. If $\alpha_\sigma(s)$ is the face of Δ_n not containing ν_k and $\alpha_\tau(t)$ is the face not containing ν_ℓ, then $f_{i(k)} = f_{j(\ell)}$.
3. $(-1)^{k+\ell}\lambda_i\lambda_j < 0$.

The last condition ensures that the contribution of the faces s and t in c cancel out when the boundary is taken.

From Y a manifold with singularities is built up by identifying $(n-1)$-*simplexes*. Pick two identifiable pairs (σ, s) and (τ, t) in Y and using the notation above identify s and t by the rule $x \sim y$ if $\partial_k^{-1}\alpha_\sigma(x) = \partial_\ell^{-1}\alpha_\tau(y)$. This reduces the number of identifiable pairs by one. Continue in this fashion until none is left. The resulting complex K is a circuit. Let $f : K \longrightarrow X$ be defined on n-*cells* by the compositions $f_i\alpha_\sigma$. It is not hard to see that $f : K, \partial K \longrightarrow X, A$ corresponds to the conditions of the theorem.

1.4 Cohomology.

A cohomology theory, dual to homology, may be defined using the Eilenberg-Steenrod axioms but with the arrows reversed. If homology with $\mathbb{Z}$ coefficients is calculated using the chain complex $\{C_n, \partial\}$ then cohomology with coefficients in A can be calculated from the cochain complex $\{C^n, \delta\}$ given by

$$C^n = Hom(C_n, A) \quad \text{and} \quad \delta : C^n \longrightarrow C^{n+1}$$

defined by the rule $(\delta h)(c) = h(\partial c)$. As with homology, we call elements of $Ker\{C^n \longrightarrow C^{n+1}\}$ **cocycles** and elements of $Im\{C^{n-1} \longrightarrow C^n\}$ **coboundaries**. Similarly, we talk of **cocycle** bases of H^n when they exist.

Definition 1.4.1 Kronecker duality.

Evaluation of cohomology on homology gives a pairing $H^n \otimes H_n \longrightarrow A$ called Kronecker duality and defined by $([\xi], [z]) = \xi(z)$.

1.4.2 The cohomology groups of a cell complex.

Let K be a cell complex with incidence numbers $[\sigma^i ; \sigma^{i+1}]$ defined as in 1.3.1 . Let ξ_σ be the cochain which has value 1 on σ and value 0 on every other cell. Then the coboundary homomorphism is given by the rule

$$\delta(\xi_{\sigma^i}) = \sum [\sigma^i ; \sigma^{i+1}] \xi_{\sigma^{i+1}},$$

the sum being taken over all $(i+1)$-cells σ^{i+1} in K. The relationship between the cohomology groups and the homology groups is given by the formula:

$$H^n(K; A) \approx Hom(H_n(K), A) \oplus Ext(H_{n-1}(K), A), \qquad [12].$$

Definition 1.4.3 The Betti group and torsion.

Let $H_n K = B_n \oplus T_n$ where B_n, the Betti group, is free and T_n is the torsion subgroup of H_n.

The formula above allows the following table, which gives possible values of H_n and H^n for a compact cell complex K of dimension n, to be written down.

Table 1.4.4 Comparison of homology and cohomology for a cell complex of dimension n.

i	0	1	2	...	i	...	n
H_i	B_0	$B_1 \oplus T_1$	$B_2 \oplus T_2$	...	$B_i \oplus T_i$	...	B_n
H^i	B_0	B_1	$B_2 \oplus T_1$	...	$B_i \oplus T_{i-1}$	...	$B_n \oplus T_{n-1}$

Things to note are, that the rank of H^0 and H_0 is equal to the number of components of K, that H^1 and H_n are free, that the torsion in H^i is equal to the torsion in H_{i-1}, and that the free part of H^i is equal to the free part of H_i.

Example. The cohomology of a two-dimensional complex.

If $K = \{A \mid R\}$ is a two-dimensional complex with one *0-cell* then the cohomology groups can be calculated from the homology groups and Table 1.4.4. The cochains ξ_a are cocycles if and only if $\varepsilon_a(r) = 0$ for all $r \in R$.

1.4.5 Subdivision of cell complexes.

Finally, we conclude this section with a brief consideration of subdivision.

Definition 1.4.6 Subdivision.

Let K, L be complexes with $|K| = |L|$. Then L is a **subdivision** of K if every cell of L is entirely contained within a cell of K.

Definition 1.4.7 *Sd* and *Min*.

If L is a subdivision of K let $Sd : C_p(K) \longrightarrow C_p(L)$ be defined as follows: if σ is a *p-cell* of K then it is divided into one or more *p-cells* $\{\sigma_i\}_{i\in I}$ by L. Let $Sd(\sigma) = \sum_{i\in I} \pm \sigma_i$ where the sign is chosen to be positive if σ_i is oriented compatibly with σ and negative otherwise.

Suppose that K and L are simplicial complexes. Then Sd has a left inverse called Min, and defined as follows:

Order the vertices of K. If a is a vertex of L then it lies in a unique simplex $\sigma(a)$ of K with smallest dimension, called the **carrier** of a. Let $Min(a)$ be the minimal vertex of $\sigma(a)$. This vertex assignment defines a simplicial map $L \to K$ and a homomorphism $C_p(L) \longrightarrow C_p(K)$ both denoted by Min. The fact that $Min \circ Sd = 1$ is a consequence of a generalised Sperner's lemma, see Ky Fan.

Definition 1.4.8 Chain maps and cellular maps.

A series of homomorphisms $f : C_p(K) \longrightarrow C_p(L)$ is a **chain map** if $f\partial = \partial f$.

Chain maps are the analogue of continuous maps in the topological category. They are useful because they take cycles (boundaries) into cycles(boundaries) and hence induce a homomorphism $f_* : H_p(K) \longrightarrow H_p(L)$. A continuous map $f : K \to L$ is a **cellular map** if it takes cells into cells. So if σ is a cell of K then $f\sigma$ is a cell of L with $dim\ f\sigma \leq dim\ \sigma$. Moreover, f is to preserve incidence so that

$$[f(\rho) ; f(\sigma)] = [\rho ; \sigma]$$

and if

$$[\tau ; f(\sigma)] \neq 0$$

then there is a $\rho \in K$ such that $\tau = f(\rho)$.

Any cellular map $f : K \longrightarrow L$ induces a chain map $f_{\#} : C_p(K) \longrightarrow C_p(L)$.

Theorem 1.4.9 (Invariance of homology under subdivision)

The maps Sd *and* Min *are chain maps and induce isomorphisms between* $H_p(K)$ *and* $H_p(L)$.

1.5 The cup product.

If cohomology were just a dual statement of homology, then it would be of little interest since as a group cohomology can be calculated from homology with integer coefficients. However, a product can be defined on cohomology, the cup product, which makes cohomology into a graded ring, an altogether more powerful and interesting object. In many ways the cup product is the 'soul' inside all topological existence theorems. Statements like: two lines in the projective plane meet; a real continuous function takes all intervening values; and so on, are all manifestations of the cup product.

Definition 1.5.1 Theoretical definition of the cup and cap products.

Let K and L be complexes and $K \times L$ the product complex. If f, g are cochains on K and L respectively, a cochain $f \times g$ can be defined on $K \times L$ by the formula:

$$(f \times g)(\sigma_1 \times \sigma_2) = f(\sigma_1)\, g(\sigma_2) .$$

This is the **cross product** for cohomology and induces a pairing:

$$H^p(X) \otimes H^q(L) \longrightarrow H^{p+q}(K \times L) .$$

Let $d : K \longrightarrow K \times L$ be the **diagonal map**

$$d(x) = (x, x), \qquad [13] .$$

The **cup product** is given by the composition

$$H^p(K) \otimes H^q(K) \longrightarrow H^{p+q}(K \times K) \xrightarrow{d^*} H^{p+q}(K) .$$

The cup product of two classes u and v is written $u \cup v$ or uv. Notice that it is necessary to have a contravariant functor, so that the homomorphism d^* goes the right (wrong?) way.

The cup product is **associative** $(u \cup v) \cup w = u \cup (v \cup w)$ and **anti-commutative** $u \cup v = (-1)^{\dim u \dim v}\, v \cup u$; moreover, there is a unit class 1 which is the image of the unit under the induced homomorphism $H^0(point) \longrightarrow H^0(K)$.

The **cap product** pairing $H^p(K) \otimes H_n(K) \longrightarrow H_{n-p}(K)$, is given by the rule $(u, v \cap x) = (u \cup v, x)$.

1.5.2 Cochain products.

In order to define cup and cap products on the chain and cochain level, it is necessary to define a cellular approximation to the diagonal map d. This was provided by Whitney in 1938, where K is a regular cell complex.

Definition 1.5.3 Regular cell complexes.

A **regular cell complex** is a complex in which all the cells σ are embedded and their boundaries $\partial\sigma$ are subcomplexes.

Unfortunately, regular cell complexes are very wasteful in

the amount of cells required. For example, a regular cell complex decomposition of the torus requires at least 16 cells, whereas the minimal decomposition only requires 4. Consequently, Whitney's chain approximation is not a great deal of use for calculations, even though it is important theoretically. For this reason, we only give Whitney's rule in the simplest case: that is, when K is a simplicial complex.

Definition 1.5.4 **The Whitney chain map.**

Let K be a simplicial complex in which the vertices are ordered. So any *n-simplex* σ in K may be written uniquely as an ordered *(n+1)-tuple* of vertices

$$\sigma = a_0 a_1 \ldots a_n, \qquad a_0 < a_1 < \ldots < a_n.$$

Define the **front p-face** of σ to be

$$_p\sigma = a_0 a_1 \ldots a_p, \qquad 0 \le p \le n.$$

Let the **back q-face** of σ be defined by

$$\sigma_q = a_{n-q}\, a_{n-q+1} \ldots a_n, \qquad 0 \le q \le n.$$

If σ, τ are simplexes of K such that the last vertex of σ is the first vertex of τ and such that the union of their vertices spans a simplex of K then write $\langle \sigma\tau \rangle$ for this simplex.

So σ may be written $\sigma = \langle {}_p\sigma\, \sigma_q \rangle$ if $p+q = n$.
Let $\xi \in C^p$ and $\eta \in C^q$. Define $\xi \cup \eta \in C^{p+q}$ by

$$(\xi \cup \eta)(\sigma) = \xi({}_p\sigma) \,.\, \eta(\sigma_q), \qquad [14].$$

Theorem 1.5.5

The definition of $\cup$ *above is a cochain approximation to the cup product on cohomology.*

(This means that if ξ and η are cocycles then so is $\xi \cup \eta$ and $[\xi \cup \eta] = [\xi] \cup [\eta]$)

The proof of 1.5.5 will be delayed until a seemingly irrelevant subdivision of an *n-simplex* is defined.

Definition 1.5.6 The prismatic subdivision of a simplex.

Let σ be a simplex with vertices $a_{00}, a_{11}, \ldots, a_{nn}$ and let a_{ij} be the barycentre of the points a_{ii} and a_{jj}, $i \leq j$. The vertices of the **prismatic subdivision** $P(\sigma)$ are the $(n+1)(n+2)/2$ points a_{ij}, $0 \leq i \leq j \leq n$. The cells of the subdivision are called **prism cells.** A collection of pq vertices $\{a_{ij}\}$, $i = i_1, \ldots, i_p$, $j = j_1, \ldots, j_q$ are to span a prism cell in $P(\sigma)$ if and only if $i_1 < i_2 < \ldots < i_p = j_1 < j_2 < \ldots < j_q$.

Lemma 1.5.7 *The prism cells subdivide* σ *into a cell complex.*

Proof. Assume inductively that $\partial\sigma$ is already subdivided. Any point x in the interior of σ can be written as an affine sum $x = \sum_{i=0}^{n} \lambda_i a_{ii}$ where $\sum_{i=0}^{n} \lambda_i = 1$ and $0 < \lambda_i < 1$.

There is a unique r such that either:

1. $\lambda_r + \lambda_{r+1} + \ldots + \lambda_n = \frac{1}{2}$.

or 2. $\lambda_r + \lambda_{r+1} + \ldots + \lambda_n > \frac{1}{2}$ *and* $\lambda_{r+1} + \lambda_{r+2} + \ldots + \lambda_n < \frac{1}{2}$.

In both cases the point lies in the prism cell spanned by the vertices $\{a_{ij}\}$, $0 \leq i \leq r$, $r \leq j \leq n$. In the first case the point lies in the boundary, in the second case the point lies in the interior of the prism cell.

This shows that σ is the union of the prism cells. The interiors of the prism cells will be disjoint, because of the uniqueness of the expression above.

An illustration of the subdivision for $n = 3$ is given in Figure 4.

Definition 1.5.8 The Whitney chain map.

If K is a simplicial complex, let $d_\# : C_n(K) \longrightarrow C_n(K \times K)$ be defined on *n-simplexes* by

$$d_\# (\sigma) = \sum_{p+q=n} {}_p\sigma \times \sigma_q .$$

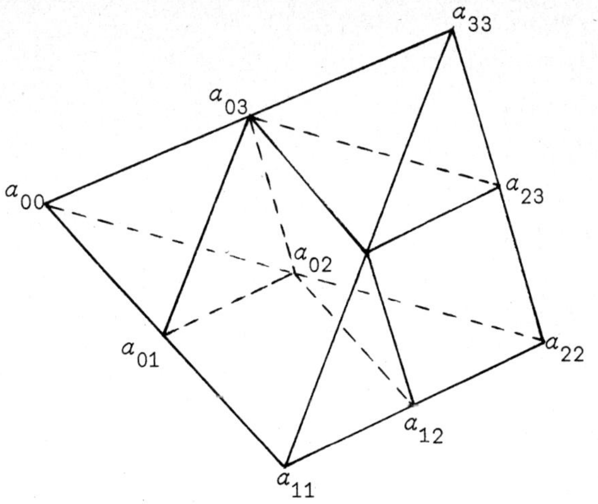

Figure 4. The prismatic subdivision of a 3-simplex into two triangular prisms and two tetrahedra.

Theorem 1.5.9 *The map* $d_\#$ *is a chain map and induces the same homomorphism on homology as the diagonal map* d.

Proof. The fact that $d_\#$ is a chain map, $d_\# \partial = \partial d_\#$, can be verified by a straightforward calculation.

Now subdivide K into $P(K)$ by the prism subdivision. Take a typical simplex σ in K and suppose that the vertices of the prism subdivision of a simplex are $\{a_{ij}\}$, $0 \le i \le j \le n$, as considered in the previous section.

Define the cellular map $\Delta : P(K) \longrightarrow K \times K$ by $\Delta(a_{ij}) = a_i \times a_j$. Then it is easy to see that the composition $K \xrightarrow{Sd} P(K) \xrightarrow{\Delta} K \times K$ induces $d_\# : C_n(K) \longrightarrow C_n(K \times K)$ on the chain level. So the result will follow if it can be shown that $\Delta \circ Sd$ is homotopic to the diagonal map.

Subdivide, by the prism subdivision, the simplex $d(\sigma)$ with vertices $a_i \times a_j$, $i = 0, \ldots, n$. Let the vertices be b_{ij}, $0 \leq i \leq j \leq n$ and let $\gamma_{ij}(t)$ be the linear path in $\sigma \times \sigma$ such that

$$\gamma_{ij}(t) = \begin{cases} a_i \times a_j & \text{if } t = 0 \\ b_{ij} & \text{if } t = 1. \end{cases}$$

Of course $\gamma_{ij}(t)$ will always be the point $a_i \times a_i$ if $i = j$.

Then the paths γ_{ij} define a homotopy of semilinear maps joining $\Delta \circ Sd$ and the diagonal.

The proof of 1.5.3 now follows, by noting that the formula for $\xi \cup \eta$ is dual to the Whitney chain map in the sense that

$$(\xi \cup \eta)(\sigma) = (\xi \times \eta)\, d_\#(\sigma) = \xi({}_p\sigma) \cdot \eta(\sigma_q), \quad p = \dim \xi, \quad q = \dim \eta,$$

all other products vanishing.

1.5.10 Examples.

1. Product complexes $K \times L$

The appropriate diagonal chain approximation is

$$d_\#(\sigma \times \tau) = \sum (-1)^{qr}\, {}_p\sigma \times {}_q\tau \times \sigma_r \times \tau_s,$$

the sum taken over all p, q, r, s such that $p + r = \dim \sigma$ and $q + s = \dim \tau$. This leads to the formula

$$(\xi \times \eta) \cup (\alpha \times \beta) = (-1)^{\dim \eta \dim \alpha} (\xi \cup \alpha) \times (\eta \cup \beta).$$

Let $X = S^p \times S^q$ then $H^{p+q}(X)$ is generated by $u \times 1$ and $1 \times v$ where u generates $H^p(S^p)$ and v generates $H^q(S^q)$. By the formula above, $(u \times 1) \cup (1 \times v) = u \times v$. On the other hand

$$(u \times 1) \cup (u \times 1) = 0 \text{ and}$$
$$(1 \times v) \cup (1 \times v) = 0.$$

2. The Mickey Mouse space. [15].

The wedge product $S^2 \vee S^1 \vee S^1$ has the same homology and cohomology groups as the torus $S^1 \times S^1$. However, all cohomology products in $S^2 \vee S^1 \vee S^1$ vanish, as the reader may readily verify using the cochain formula.

1.6 The homology and cohomology of Manifolds.

For manifolds, there is a natural isomorphism between homology and cohomology. Under this correspondence the cup product becomes an easily visualised product on homology - the intersection or dot product. Consequently, manifolds are one of the easiest spaces for calculating the cup product.

For simplicity, all manifolds considered in this book will be assumed to have a combinatorial structure [15].

Definition 1.6.1 Combinatorial manifolds.

A manifold M is said to be **combinatorial** if for one (and hence every) triangulation K, the link of an *i-simplex* is an *(n - i - 1)-sphere* or an *(n - i - 1)-ball*; the latter simplexes belonging to the boundary ∂M.

Definition 1.6.2 Orientation, etc.

We have talked previously in a rather glib manner about orientation of simplexes etc. Unfortunately, the precise meaning of orientation is complicated. The easiest method of attack is to use local homology groups.

Let M be an *n-manifold* and B an *n-ball* embedded in M. Then a local **orientation** defined on B is a choice of generator of $H_n(M, M - B) \approx H_n(B, \partial B) \approx \mathbb{Z}$, $n > 0$. If $n > 1$ then $H_n(B, \partial B) \approx H_{n-1}(\partial B)$. So a choice of generator of $H_{n-1}(\partial B)$ determines a choice of generator of $H_n(B, \partial B)$. To see how this works out in practice, consider the standard orientations of the standard simplexes.

If $n = 0$, a generator of $H_0(point)$ is represented by the point itself and will be indicated by a + sign. The opposite orientation is of course indicated by a - sign.

If $n = 1$, a generator of $H_1(\Delta_1, \partial\Delta_1)$ is chosen by the *1-simplex* with boundary $v_1 - v_0$.

If $n > 1$, the standard orientation of Δ_n is chosen inductively by the rule $[\Delta_{n-1}; \Delta_n] = +1$. In other words, the orientation of Δ_{n-1} orients $\partial\Delta_n$ and hence Δ_n.

This definition may differ from that given by other writers.

The orientation of a general simplex is determined by a homeomorphism of the standard simplex onto it, which in turn may be defined by an ordering of its vertices. Two orderings determine the same orientation if and only if they differ by an even permutation.

The orientation of a *1-simplex* is designated by an arrow, $- \bullet\!\!\longrightarrow\!\!\bullet +$. The orientation of a *2-simplex* is determined by the orientation of its boundary circle and is designated by a curved arrow ↻. If B_1 is an *n-ball* inside an *n-ball* B_2 then an orientation of B_1 induces an orientation on B_2 and conversely. These orientations are called **compatible**.

Definition 1.6.3 Orientable manifolds.

An *n-manifold* M is **orientable** if every *n-ball* in M can be given a mutually compatible orientation. Otherwise, the manifold is **non-orientable**.

If B_1, B_2 are two *n-balls* and $B_1 \cap B_2 = \partial B_1 \cap \partial B_2$ is an *(n - 1)-ball* then $B_1 \cup B_2$ is an *n-ball* and B_1, B_2 are said to be **adjacent**.

Their orientations are said to be **coherent** if they induce opposite orientations in the *(n - 1)-ball* $B_1 \cap B_2$.

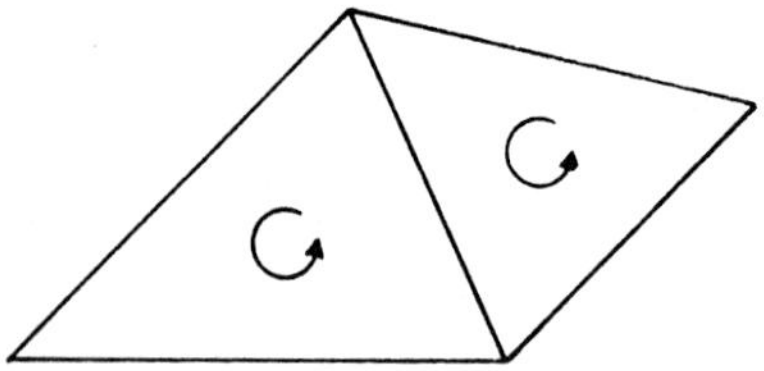

Figure 5. Coherent orientations on adjacent 2-simplexes.

Let K be a triangulation of a compact orientable *n-manifold* M. Then the *n-simplexes* of K can be oriented coherently so that their sum is a cycle in $K, \partial K$. If M is connected this class $[K]$ generates the infinite cyclic group $H_n(M, \partial M)$. If M is non-orientable the sum of all the *n-simplexes* of K is still a cycle of $K, \partial K$ over the integers *mod 2*.

Theorem 1.6.4 Duality.

The cap product $u \longrightarrow u \cap [K]$ *induces a natural isomorphism* $H^p(M) \longrightarrow H_{n-p}(M, \partial M)$. *($\mathbb{Z}$ coefficients if M is orientable and $\mathbb{Z}_2$ coefficients otherwise.)*

The above result is usually called **Poincaré duality** if $\partial M = \emptyset$, and **Lefschetz duality** if $\partial M \neq \emptyset$.

Definition 1.6.5 Codimension.

If L is a subcomplex of the manifold K, $\dim K = n$, $\dim L = p$, then $n-p$ is called the **codimension** of L. So duality induces an isomorphism between homology classes of dimension p and cohomology classes of codimension p.

This allows the following tables to be written down for the possible values of homology and cohomology in an orientable connected closed compact manifold of dimension n. There are two cases, according to whether n is even or odd.

Table 1.6.6 Homology for an orientable manifold of dimension $n = 2k$.

p	0	1	... p	... $k-1$	k	$k+1$
H_p	$\mathbb{Z}$	$B_1 \oplus T_1$	... $B_p \oplus T_p$	... $B_{k-1} \oplus T_{k-1}$	$B_k \oplus T_{k-1}$	$B_{k-1} \oplus T_{k-2}$
H^p	$\mathbb{Z}$	B_1	... $B_p \oplus T_{p-1}$	... $B_{k-1} \oplus T_{k-2}$	$B_k \oplus T_{k-1}$	$B_{k-1} \oplus T_{k-1}$

... $2k-p$	... $2k-1$	$2k$
... $B_p \oplus T_{p-1}$	... B_1	$\mathbb{Z}$
... $B_p \oplus T_p$	... $B_1 \oplus T_1$	$\mathbb{Z}$

Table 1.6.7 Homology for an orientable manifold of dimension $n = 2k+1$.

p	0	$1 \; \ldots$	p	$\ldots$	k	$k+1$	$\ldots$	$2k+1-p$	$\ldots$	$2k$	$2k+1$
H_p	$\mathbb{Z}$	$B_1 \oplus T_1 \; \ldots$	$B_p \oplus T_p$	$\ldots$	$B_k \oplus T_k$	$B_k \oplus T_{k-1}$	$\ldots$	$B_p \oplus T_{p-1}$	$\ldots$	B_1	$\mathbb{Z}$
H^p	$\mathbb{Z}$	$B_1 \; \ldots$	$B_p \oplus T_{p-1}$	$\ldots$	$B_k \oplus T_{k-1}$	$B_k \oplus T_k$	$\ldots$	$B_p \oplus T_p$	$\ldots$	$B_1 \oplus T_1$	$\mathbb{Z}$

Proof of 1.6.4

Duality is proved on the chain level. Since homology and cohomology are topological invariants, we are allowed to use whatever cell decomposition is convenient. For cohomology, this is an ordinary simplicial complex K triangulating M. For homology, a dual cell subdivision is used. The idea goes back at least to Cauchy.

Definition 1.6.8 The barycentric subdivision K' and the dual cell decomposition K^*.

If σ, τ are simplexes of K write $\tau \prec \sigma$ if τ is a proper face of σ; (so $\tau \neq \emptyset$ and $\tau \neq \sigma$). Let $\hat{\sigma}$ be the barycentre of σ. Considering σ as the affine hull of its vertices $a_0, a_1, \ldots, a_p$ in some vector space, then

$$\hat{\sigma} = \frac{1}{p+1}\left\{a_0 + a_1 + \ldots + a_p\right\}.$$

The vertices of K' are the barycentres $\hat{\sigma}$ as σ varies over K. So there is a *1-1 correspondence* $\sigma \to \hat{\sigma}$ between the simplexes of K and the vertices of K'. The vertices $\hat{\sigma}_0, \hat{\sigma}_1, \ldots, \hat{\sigma}_p$ are to span a simplex of K' if and only if $\sigma_0 \prec \sigma_1 \prec \ldots \prec \sigma_p$ in K.

Notice that K' has an ordering $\hat{\sigma} < \hat{\tau}$ defined on its vertices whenever $\sigma \prec \tau$. If σ is a *p-simplex* of K then σ is subdivided into the set of all simplexes of K' of the form $a_0 a_1 \ldots a_q$ where the right-hand member $a_q = \hat{\sigma}$. Dually the cell σ^* is the union of all simplexes $b_0 b_1 \ldots b_r$ where the left-hand member $b_0 = \hat{\sigma}$. To see that σ^* is an *$(n-p)$-cell*, note that it is a join $\sigma^* = \hat{\sigma} L(\sigma)$ where $L(\sigma)$ is homeomorphic to the link of σ in K.

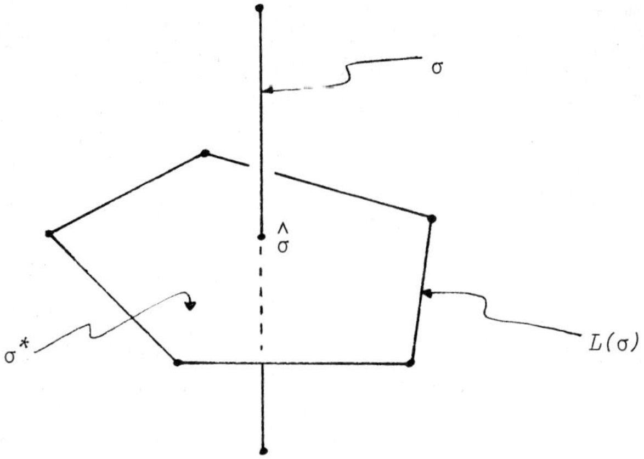

Figure 6. A simplex σ and its dual cell σ^*.

Definition 1.6.9 The orientation of the dual cells σ^*.

Assume that the simplexes of K are oriented. The dual cell σ^* is transverse to σ and can therefore be oriented by the following rule:

If $dim\ \sigma = 0$ then $dim\ \sigma^* = n$ and so inherits an orientation from the manifold.

If $dim\ \sigma > 0$ define orientation inductively by the rule:

Definition 1.6.10

$$[\sigma^* ; \tau^*] = (-1)^p [\tau ; \sigma], \text{ where } p = dim\ \sigma .$$

Definition 1.6.11 The duality homomorphism.

The correspondence $\sigma \to \sigma^*$ induces an isomorphism $C^i(K) \xrightarrow{\psi} C_{n-i}(K^*)$. By 1.6.10 ψ takes cocycles into cycles $mod\ \partial M$. So ψ induces an isomorphism M between cohomology and homology. It only remains to show that the induced isomorphism is given by $\cap [K]$.

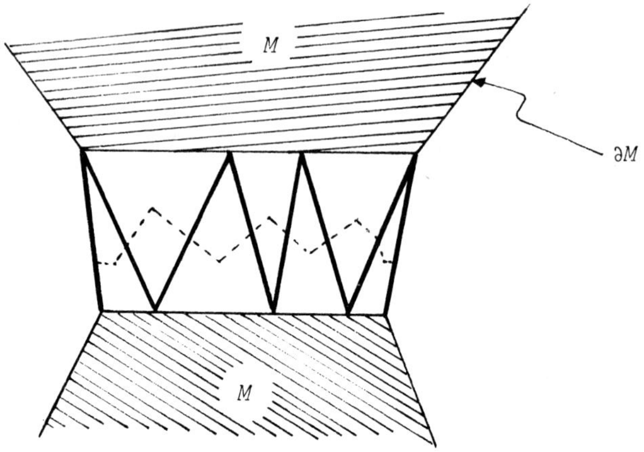

Figure 7. A cocycle in K (heavy lines) and its dual cycle in K^* (dotted line).
Notice the resemblance between a cocycle and a disc bundle over the dual cycle.

Let $Min : C_p(K') \to C_p(K)$ be the chain map which sends each vertex of K' to the minimal vertex of its carrier in K. Then Min induces a natural isomorphism between the homology of K' and the homology of K.

If ξ is a cocycle in $C^p(K)$ then $Min\ \xi^*$ is a cycle in $C_{n-p}(K, \partial K)$. We shall show that this is the same as the cycle z given by the formula $z = \sum \lambda_\sigma \sigma$ where $\lambda_\sigma = (\xi_\sigma, \xi \cap K) = (\xi_\sigma \cup \xi, K)$.

Now λ_σ is a sum of elements of the form $\xi(\tau)(\xi_\sigma \cup \xi_\tau, K)$ and the evaluation $(\xi_\sigma \cup \xi_\tau, K)$ is non-zero if and only if there is an *n-simplex* $\langle \sigma\tau \rangle$, such that ${}_{n-p}\langle \sigma\tau \rangle = \sigma$ and $\langle \sigma\tau \rangle_p = \tau$.

On the other hand, if τ is a *p-face* of the *n-simplex* ρ then $Min(\tau^* \cap \rho)$ is a non-degenerate $(n-p)$*-face* σ of ρ if and only if the smallest vertex of τ is the largest vertex of σ, i.e. $\rho = \langle \sigma\tau \rangle$. The result now follows, since $\xi^* = \sum \xi(\tau)\tau^*$.

Lemma 1.6.12.

Suppose σ, τ *are simplexes of* K *with* $\dim\sigma \le \dim\tau$. *Then* $\sigma^* \cap \tau$ *is non-empty if and only if* σ *is a face of* τ *and in that case* σ^* *meets* τ *transversely in a cell of dimension* $\dim\tau - \dim\sigma$.

Proof. A simplex of K' in $\sigma^* \cap \tau$ is of the form $a_0 \ldots a_k$ where $a_0 = \hat{\sigma}$ and $a_k = \hat{\tau}$. This is non-empty if and only if either $\sigma = \tau$ or there is a sequence $\sigma = \sigma_0 \prec \sigma_1 \prec \ldots \prec \sigma_k = \tau$. Either way, σ must be a face of τ. If this is the case, $\sigma^* \cap \tau$ is the dual of σ in the complex defined by τ and its faces and hence is a cell with the required properties.

Definition 1.6.13 The intersection product.

Since σ and τ are oriented, the intersection $\sigma^* \cap \tau$, if non-empty, can be oriented in the usual manner. Let $\sigma^* \cdot \tau$ be the sum of the correspondingly oriented cells of K' lying in $\sigma^* \cap \tau$ of dimension $\dim\tau - \dim\sigma$.

This defines the intersection product

$$C_p(K^*) \otimes C_q(K) \xrightarrow{\ \cdot\ } C_{p+q-n}(K').$$

By checking orientations, this product is seen to satisfy the rule:

$$\partial(c \cdot d) = \varepsilon\, \partial c \cdot d + c \cdot \partial d,$$

where $$\varepsilon = (-1)^{\dim d}.$$

Hence if c and d are cycles (boundaries) then $c \cdot d$ is also a cycle (boundary) and the intersection product carries over to a pairing

$$H_p(M, \partial M) \otimes H_q(M) \longrightarrow H_{p+q-n}(M) .$$

Theorem 1.6.14

The intersection product on manifolds is dual to the cup product on manifolds.

Proof. If $p+q = n$ then the result follows by duality.

If $p+q \geq n$ then we use an indirect argument due to Whitney.

Firstly, it is convenient to consider the associated cap product

$$C^p(K) \otimes C_q(K) \xrightarrow{\ \cap\ } C_{q-p}(K') ,$$

given by $\xi \cap c = \xi^* \cdot c$ where $\xi \in C^p(K)$ and $c \in C_q(K)$.

Secondly, we consider all the properties which we might reasonably expect a cap product intersection to satisfy. Call the products satisfying these conditions natural.

Then we show that the above cap product is natural.

Finally, we show that all natural products give rise to the same product when homology and cohomology classes are taken.

Now the sort of properties which we might expect a bilinear pairing

$$C^p(K) \otimes C_q(K) \xrightarrow{\ \circ\ } C_{q-p}(K')$$

to satisfy are of three types:

1. Locality. We should expect that the product of a cochain with a simplex τ far away to be zero. So we require that the support of the chain $\xi \circ \tau$ lie in the support of ξ lying in τ

2. Identity property. We should expect that the product of a dual simplex with the same simplex should be the identity in some sense. So, if ξ_σ is as usual the cochain dual to the simplex σ we require that $\xi_\sigma \circ \sigma$ be a vertex of K' in σ .

3. **Derivation property.** Since the product ought to respect homology and cohomology, we would expect it to satisfy the usual derivation formula of the form

$$\partial(\xi \circ c) = (-1)^{p-q} \delta\xi \circ c + \xi \circ \partial c$$

for $\xi \in C^p(K)$ and $c \in C_q(K)$.

It is now a straightforward task to verify that $\cap$ given above is a natural product.

Let $\overset{1}{\circ}$ and $\overset{2}{\circ}$ be two such natural products. We now define a bilinear pairing $\xi \wedge c$ satisfying:

If $\xi \in C^p(K)$, $c \in C_q(K)$, then

1. $\xi \wedge c$ is a $(q-p+1)$-*chain* in K' with support in c.
2. If $p = q$ then $\xi \overset{1}{\circ} c - \xi \overset{2}{\circ} c = \partial(\xi \wedge c)$.
3. If $p < q$ then $\xi \overset{1}{\circ} c - \xi \overset{2}{\circ} c = \partial(\xi \wedge c) + (-1)^{q-p} \delta\xi \wedge c + \xi \wedge \partial c$.

The pairing $\wedge$ is constructed on simplexes and then extended by linearity.

Now $\xi_\sigma \overset{1}{\circ} \sigma - \xi_\sigma \overset{2}{\circ} \sigma$ is the difference of two vertices in σ. So there is a *1-chain* $\xi_\sigma \wedge \sigma$ in K' of the required form.

Suppose that $\xi \wedge c$ is defined for all $q-p < r$ and let $q-p = r$.

Put $x = \xi \overset{1}{\circ} c - \xi \overset{2}{\circ} c - (-1)^{q-p} \delta\xi \wedge c - \xi \wedge \partial c$.

Then x is a cycle and if c is a simplex σ then x has support in σ. So x is a boundary.

Let $\xi \wedge c$ satisfy $x = \partial(\xi \wedge c)$. Using $\xi \wedge c$ and the three properties above, we see that $\overset{1}{\circ}$ and $\overset{2}{\circ}$ induce the same product when homology and cohomology classes are taken.

This means that for manifolds at least the cup product of two cohomology classes can be calculated, as follows: represent the classes by cocycles and dualise so that they are represented cycles in K^*. Move one of the cycles so that they meet transversely. This can be achieved by subdividing one cycle so that it is a cycle in K' and then applying Min so that it is a cycle in K. (If the triangulation K has fine enough mesh the cycle will be altered by an arbitrarily small amount.) Now take their intersection, which will be a cycle in K'. Its dual class will represent the cup product of the original cohomology classes.

In Section 1.8 this method will be generalised to arbitrary polyhedra.

Example:

Let T_γ be the closed orientable surface of genus γ and let $a_1, b_1, a_2, b_2, \ldots, a_\gamma, b_\gamma$ be the cycles shown in Figure 6. These form a cycle basis of $H_1(T_\gamma)$.

If $(x_1, x_2, x_3, \ldots, x_{2\gamma}) = (a_1, b_1, a_2, \ldots, b_\gamma)$ then the matrix $x_i \cdot x_j$ is the direct sum of the matrices

$$\begin{pmatrix} 0 & 1 \\ -1 & 0 \end{pmatrix}.$$

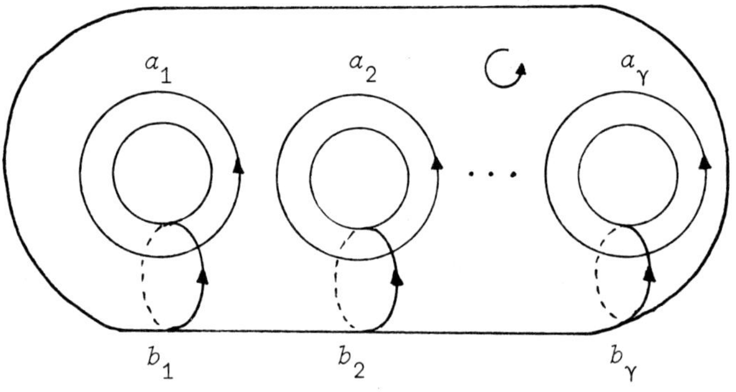

Figure 6. A cycle basis for $H_1(T_\gamma)$.

1.7 Non-orientable manifolds.

Although non-orientable manifolds only satisfy Poincaré duality with respect to $\mathbb{Z}_2$ coefficients, some things can still be said about ordinary homology with $\mathbb{Z}$ coefficients. The duality between a triangulation and its dual still works, but care must be taken with the sign if the cochain or chain contains a loop which reverses orientation.

Theorem 1.7.1 *Let M be a closed compact connected non-orientable manifold of dimension n, then:*

1. $H_n(M) = 0$.
2. $H^n(M) \approx \mathbb{Z}_2$.

Proof.
To prove 1. is easy, since by definition no set of *n-simplexes* can be coherently oriented to form a non-trivial cycle.
To prove 2. Let K be a triangulation of M and as usual let ξ_σ be the *n-cochain* dual to an *n-simplex* σ of K. Then ξ_σ is a cocycle and determines a class $[\xi_\sigma]$ in $H^n(M)$. Let τ be another *n-simplex*. The dual cells $\hat{\sigma}$ and $\hat{\tau}$ can be joined by a *1-chain* c^* in K^*, i.e. $\partial c^* = \hat{\sigma} - \hat{\tau}$. The dual cochain c has the property that $\delta c = \xi_\sigma \pm \xi_\tau$. By a careful choice we can assume that $\delta c = \xi_\sigma - \xi_\tau$. So $[\xi_\sigma]$ is independent of σ.

Clearly the cocycles ξ_σ generate $C^n(K)$ and hence $H^n(K)$ is cyclic.

Let d^* be a loop in K^* which starts and finishes at $\hat{\sigma}$ but reverses orientation. Then $\delta d = 2\xi_\sigma$. It only remains to show that ξ_σ is not homologous to zero, but this follows from the general principal that *1-chains* cannot have just one end (remember $\partial M = \phi$).

Definition 1.7.2 **The fundamental class.**

Since $H^n(M)$ is isomorphic to $\mathbb{Z}_2$ it follows that T_{n-1} the torsion in H_{n-1}, is also isomorphic to $\mathbb{Z}_2$. A cycle representative of the generator can be described as follows.

Orient the *n-simplexes* $\{\sigma_i\}_{i=1}^{k}$ in some arbitrary fashion and let $b = \sigma_1 + \sigma_2 + \ldots + \sigma_k$. An *(n-1)-simplex* in K either contributed nothing or two to the boundary cycle ∂b; the contribution depending on whether the adjacent *n-simplexes* are oriented coherently or not.

So $\partial b = 2z$ for some *(n-1)-cycle* z. The class of z is the fundamental class of M.

If x is a loop in K^* which reverses orientation, then $x \cdot z$ is an odd number of points.

So $M - |z|$ is orientable.

1.7.3 Examples: non-orientable surfaces.

Consider firstly the projective plane: $H^1(P; \mathbb{Z}_2)$ is $\mathbb{Z}_2$. A cocycle representative ξ can be taken as the dual to a projective line. Since two distinct projectivities meet in a point, $\xi \cup \xi$ is the generator of $H^2(P; \mathbb{Z}_2)$ which is also $\mathbb{Z}_2$.

Secondly, consider the Klein bottle K; let α and β be $\mathbb{Z}_2$-*cocycles* dual to the cycles a and b shown in Figure 7.

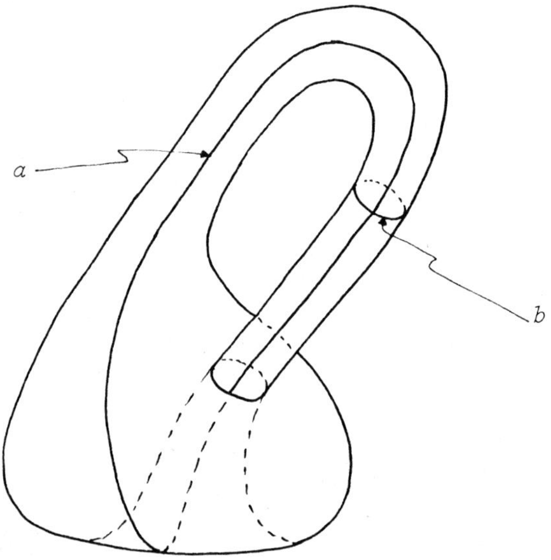

Figure 7. A cycle basis of $H_1(K; \mathbb{Z}_2)$.

Then the cup product over $\mathbb{Z}_2$ is given by the matrix

$$\begin{array}{c c} & \begin{array}{cc} \alpha & \beta \end{array} \\ \begin{array}{c} \alpha \\ \beta \end{array} & \begin{pmatrix} 1 & 1 \\ 1 & 0 \end{pmatrix} \end{array} .$$

With $\mathbb{Z}$ coefficients $H^1(K)$ is cyclic and a cocycle representative is dual to b, (with integer coefficients $\delta\alpha \neq 0$).

The class of the cup product square is zero.

Finally, a non-orientable surface is the connected sum of an orientable surface and either a projective plane or a Klein bottle. The cohomology ring is the direct sum of the orientable case and one of the two cases considered above.

1.8 Geometric Cohomology.

This last section is an attempt to generalise the geometric interpretation of cup products in a manifold (intersection products) to cup products in a general complex. This is possible because complexes are built up from cells which are themselves manifolds, at least in their interiors. The cup product will now be represented by intersection within the interior of these cells.

The idea goes back to Whitney in 1947 and has since been used by Thom, Sullivan and others. The present approach is based on the idea of a mock bundle, invented by Buoucristiano, Rourke and Sanderson.

Definition 1.8.1 Geometric cocycles.

Let K be a regular cell complex. A **geometric cocycle** ξ/K over K consists of:

1. a simplicial subdivision K_ξ of K;
2. a cocycle ξ in $C^i(K_\xi)$.

The dimension of ξ/K is defined to be i. If σ is an *n-cell* of K let $K_\xi(\sigma)$ be the subcomplex of K_ξ which subdivides and let $K_\xi^*(\sigma)$ be the dual cell complex. Given an *i-simplex* τ in $K_\xi(\sigma)$ there is a unique *(n-i)-cell* τ^* in $K_\xi^*(\sigma)$ dual to τ and conversely. Of course τ^* depends on the ambient cell σ.

If L is a collection of K cells (not necessarily forming a complex) let

$$K_\xi(L) = \bigcup_{\sigma \in L} K_\xi(\sigma).$$

Let $\xi | L$ denote the restriction of ξ to $K_\xi(L)$. Since each cell in K has a built-in orientation, there is associated with each $\xi | \sigma$ a dual chain $c_\xi(\sigma)$ in $C_{n-i}(K_\xi^*(\sigma))$. In a similar vein, let

$$c_\xi(L) = \sum_{\sigma \in L} c_\xi(\sigma).$$

The boundary cycle $\partial c_\xi(\sigma)$ lies in $C_{n-i-1}(K_\xi^*(\partial\sigma))$, and if the boundary cells of σ are oriented compatibly with σ then

$$\partial c_\xi(\sigma) = c_\xi(\partial\sigma).$$

However, at this stage we are not too concerned about the orientations of the chains $c_\xi(\sigma)$. This will be more important when we consider the interpretation of cap products later. For example, consider the example shown in Figure 8. This shows chains $c_\xi(\sigma)$ with a symbolic transverse orientation induced by the cocycle ξ. However, no orientations which could be imposed upon the $c_\xi(\sigma)$ would make any particular geometric sense.

1.8.2 Adding geometric cocycles.

If $K_\xi = K_\eta$ and $dim\,\xi = dim\,\eta$ then ξ/K and η/K can be added in the normal way. Denote by $-\xi/K$ the cocycle with opposite orientation.

If $|K| = |L|$ and K and L have a common subdivision, then ξ/K and η/L must be subdivided before they can be added.

1.8.3 Subdivision.

Let L be a subdivision of K then the subdivision map $Sd : C_i(K) \longrightarrow C_i(L)$ induces an isomorphism on homology. The dual map $Sd^\# : C^i(L) \longrightarrow C^i(K)$ induces an isomorphism on

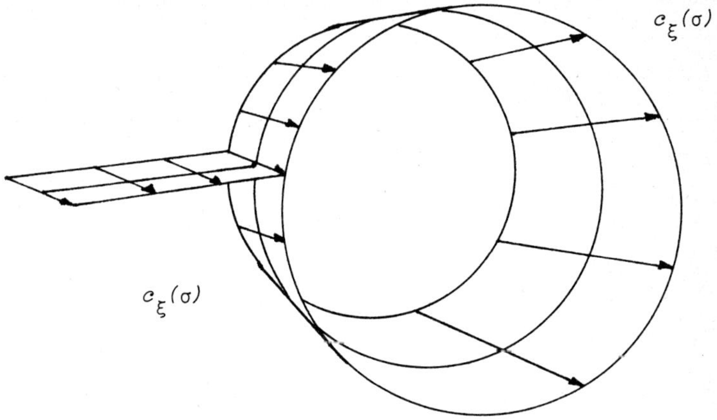

Figure 8. Geometric cochain of dimension 1 with transverse orientation.

cohomology. Now suppose that L is regular but K is not and that ξ / L is a geometric cochain. Then the definition of a geometric cochain can be extended to include irregular cell complexes by defining $Sd\,\xi / K$ with:

$$K_{Sd\xi} = L_\xi \quad \text{and} \quad Sd\,\xi = Sd^{\#}(\xi) .$$

The chains $c_{Sd\xi}(\sigma)$ are defined by

$$c_{Sd\xi}(\sigma) = c_\xi(Sd\,\sigma).$$

1.8.4 Relative cocycles.

If L is a subcomplex of K a **relative** cocycle $\xi / K, L$ defined on the pair (K, L) is a cocycle of K such that $\xi(\sigma) = 0$ for all σ in L.

1.8.5 Cobordism of cocycles : relation with cohomology.

Two *i-cocycles* ξ / K and η / L where $|K| = |L|$ are said to be cobordant if there is a subdivision J of $|K| \times I$ such that K and L are the subcomplexes at the top and bottom of J and an i-*cocycle* ω / J such that $\omega \mid K = \xi$ and $\omega \mid L = \eta$.

Associated with a cocycle ξ / K of dimension i is the cohomology class $[\xi]$ in $H^i(K_\xi)$.

Theorem 1.8.6 *Every element of* $H^i(|K|)$ *can be represented by a geometric i-cycle. Moreover, two cobordant cocycles represent the same cohomology class.*

Proof. The first part is easy : subdivide K as a simplicial complex and take a representative cocycle ξ .

Now suppose that ω is a cobordism between ξ / K and η / L, $|K| = |L|$. The inclusions

$$K \longrightarrow J \longleftarrow L \quad \text{where} \quad |J| = |K| \times I$$

are homotopy equivalences and on the cochain level represent the restrictions of ω to ξ and η .

1.8.7 **Geometric cocycles and mock bundles.**

At this stage it is interesting to compare geometric cocycles with the notion of a mock bundle given in the book by Buoucristiano, Rourke and Sanderson.

Let ξ / K be a cocycle where K is regular and consider the chains $c_\xi(\sigma)$ for $\sigma \in K$. These may be represented by geometric chains $E(\sigma) \xrightarrow{p|\sigma} \sigma$ where the $E(\sigma)$ are circuits, as follows: If $\dim \sigma = i$ then $c_\xi(\sigma)$ is a 0-*cycle* and so is represented by discrete 0-*manifolds*. Assume inductively that the $E(\tau)$ are defined for $\tau \in \partial\sigma$, $\dim \sigma > i$, then, using the techniques of 1.3.7, extend the definition to $E(\partial\sigma)$ to $E(\sigma)$ using the chain $c_\xi(\sigma)$. Let $E(\xi)$ be the union of the spaces $E(\sigma)$, $\sigma \in K$ and let p be the identification map $E(\xi) \to K$. Then $E(\sigma) = p^{-1}(\sigma)$ and $E(\partial\sigma) = p^{-1}(\partial\sigma)$. The structure obtained in this manner is like a mock bundle with 'manifold' replaced by circuit.

1.8.8 Cup products of geometric cocycles.

If ξ/K and η/K are cocycles with $K_\xi = K_\eta$ then $\xi \cup \eta$ can be defined in the usual way. However, the cup product can be 'visualised' by taking the transverse intersection of the chains c_ξ and c_η. the transverse intersections being taken inductively over $\partial\sigma$ and then extended over the interior of σ as σ varies over K.

To illustrate the preceeding discussion, consider the following:

Example. Cup products on 2-complexes.

Let $K = \{A \mid R\}$ be a *2-complex* with one *0-cell* and suppose that R has only one element, r. Let a be a member of A such that $\varepsilon_a = 0$, i.e. a has total exponent zero in the word $w(r)$. Then the cochain ξ_a dual to a is a cocycle and represents some element $u_a = [\xi_a]$ in $H^1(K)$.

Consider now the *2-cell* r attached by the word $w(r)$. This is obtained by identifying the subdivided boundary of the unit disc. The edges which are identified to a can be labelled with a^+ or a^- according to whether their appearance in the word $w(r)$ is as positive or negative exponent. These edges can be grouped into pairs of oppositely labelled exponents, since $\varepsilon_a = 0$.

Let $\hat{a}$ be the centre of a and let a_1, a_2 be the inverse image of $\hat{a}$ in a typical pair. Then a_1 and a_2 can be joined by a path α running from the negatively oriented edge to the positively oriented edge. Indeed the pairs may be chosen so that the various paths $\alpha, \beta, \ldots$ etc. are all disjoint, [19].

Now these paths form the chains $c_\xi(r)$ for some geometric cocycle ξ/K where ξ represents the class $[\xi_a]$.

If b is another generator with $\varepsilon_b = 0$ and η/K is a similar cocycle representing $[\xi_b]$ then the cup product can be calculated by looking at the transverse intersection of the chains c_ξ and c_η.

For example, in Figure 9 $[\xi \cup \eta]$ is twice a generator of $H^2(K)$.

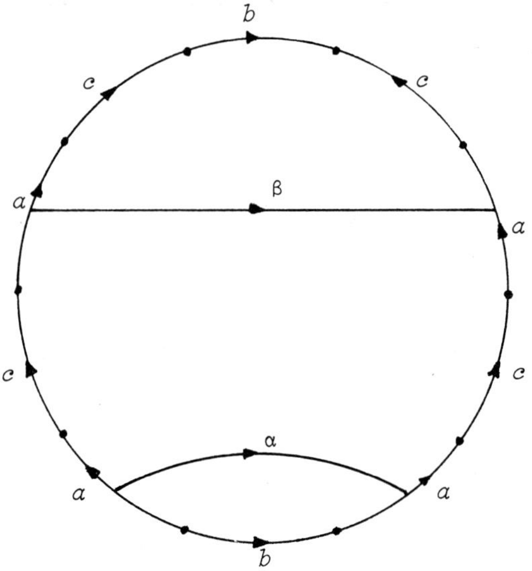

Figure 9. Two paths α, β in r when
$w(r) = b\,a\,c\,a\,c\,b^{-1}c^{-1}a^{-1}c^{-1}a^{-1} = [\,b,\ a\,c\,a\,c\,]$.

Definition 1.8.9

Let w be a word in the variables $a \in A$. Write w as

$\prod_{i=1}^{k} x_i^{\varepsilon_i}$, $x_i \in A$ and $\varepsilon_i = \pm 1$.

An occurrence of the pair a, b occurs when $x_i = a$, $x_j = b$ $i < j$. The sign of the occurrence $\ldots a^{\varepsilon_1} \ldots b^{\varepsilon_2}$ is defined to be $\varepsilon_1 \varepsilon_2$.

The integer $\varepsilon_{ab} = \varepsilon_{ab}(w)$ is the algebraic sum of the occurrences of the pair a, b, [20] .

For example,

$\varepsilon_{ab}\ (a\,b\,a^{-1}b^{-1}) = 1 - 1 + 1 = 1$

$\varepsilon_{ba}\ (a\,b\,a^{-1}b^{-1}) = -1$.

Theorem 1.8.10

With the notation above, the cup product evaluation

$([\xi_a] \cup [\xi_b], [r])$ *is* ε_{ab}, [21], *provided* $\varepsilon_c = 0$ *for all* c *in* A.

Proof. Firstly, put all generators different from a and b equal to the identity. Notice that this does not effect the value of the cup product $[\xi_a] \cup [\xi_b]$. In the new word cancel all adjacent pairs aa^{-1} etc.

Go along a path from a^{-1} to a^{+1} and note that each positive intersection with a path from b^{-1} to b^{+1} contributes $+1$ to ε_{ab} from pairs $a^{-1}b^{-1}$ or ab in $\omega(r)$ and each negative intersection contributes -1 from pairs ab^{-1} or $a^{-1}b$ in $\omega(r)$.

1.8.11 Geometric cocycles and the cap product

From the formula $(\eta, \xi \cap z) = (\eta \cup \xi, z)$ the cap product can be paraphrased as follows:

Let z be a cycle in $C_n(K)$ and let ξ/K be a cocycle of dimension i. Let σ be an n-*cell* of K. The orientation of σ induces an orientation on the dual $(n-i)$-*chain* $c_\xi(\sigma)$. If $z = \sum \lambda_\sigma \sigma$ then $\xi \cap z = \sum \lambda_\sigma c_\xi(\sigma)$ is the required cycle.

Example. The Double Banana.

Let X be a torus strangled at two meridians. Let $c = c_1 + c_2$ be a longitude and let a, b be the meridians pictured in Figure 10.

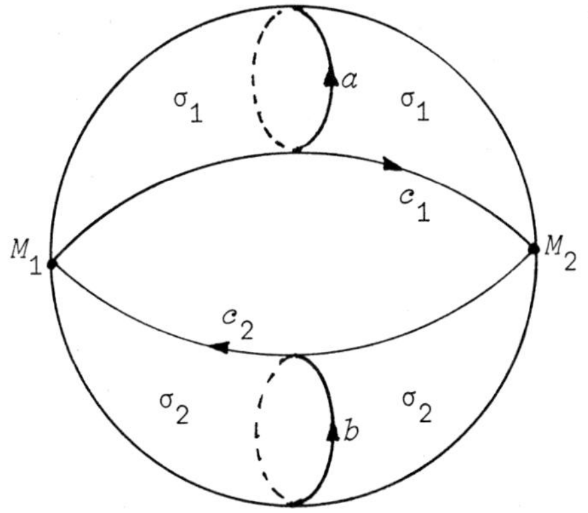

Figure 10. The double banana.

Let K be the cell sub-division of X with two vertices M_1 and M_2, two *1-cells* c_1 and c_2 and two *2-cells* σ_1 and σ_2. Let ξ/K, η/K be geometric cocycles of dimension 1 such that $K_\xi = K_\eta$ and

$$\xi(c_1) = 1 \qquad \xi(c_2) = 0$$

$$\eta(c_1) = 0 \qquad \eta(c_2) = 1$$

$$c_\xi(\sigma_1) = a \quad \text{and} \quad c_\eta(\sigma_2) = b .$$

The homology group H_1 is infinite cyclic with cycle basis $c_1 + c_2 = c$.

The cohomology group H^1 is also infinite cyclic with cocycle basis ξ or η.

To see that ξ and η are cohomologous let L be the sub-complex of K_ξ which lies to the right of a and b in Figure 10. Let f be the *0-cochain* with value 1 on all the vertices in L and value 0 on all others. Then $\delta f = \xi - \eta$.

On the other hand a and b are clearly homologous to zero but $\xi(c) = 1$ and so ξ cannot be a coboundary.

If z is the *2-cycle* $\sigma_1 + \sigma_2$ then $\xi \cap z = a$, $\eta \cap z = b$ with suitable orientations. So on the homology level $[\xi] \cap [z] = 0$.

Loosely speaking, since X is a circuit a and b can either be considered as non-trivial geometric cocycles or as trivial cycles.

1.8.12 Footnote.

One final comment on geometric cocycles may be illuminating and help the reader to understand particular examples.

Let K be a cell complex. Embed K in some R^n and let N be a regular neighbourhood. Then N is a manifold with boundary and has K as a deformation retract.

Hence any cohomology class in K may be represented by a cohomology class in M and by duality may be represented by a dual cycle in $M, \partial M$. The transverse intersection of this cycle with K corresponds to a geometric cocycle representation of the original cohomology class.

Cup products in K may be represented by transverse intersections in M and after restriction correspond to intersection in the cells of K as considered above.

Notes on Chapter 1.

1. A more general notion is a C.W. complex introduced by Whitehead in 1949. However, Listing, a student of Gauss, seems to have been the first to introduce the term complex.

2. The homotopy equivalence considered in the cell homotopy lemma is simple in Whitehead's sense. To see this, note that changing the attaching map by a, a homotopy is equivalent to an expansion followed by a collapse.

3. This method may be used to 'improve' the representative complex in the homotopy type. For example, it may be assumed that the complex is triangulable or that the attaching maps are transverse to various strata etc.

4. The topological classification of surfaces was known to Riemann in the 1850's. For *3-manifolds* no such nice result is known. However, it is known by the work of Bing and Moise that all *3-manifolds* can be decomposed as complexes. Even this is unknown for *n-manifolds*, $n \geq 4$.

5. For the reader's convenience, these are:

1. If f = identity then f_* = identity.

($f : X, A \to Y, B$) and $f_* : H_p(X, A) \to H_p(Y, B)$ is the induced map.)

2. $(gf)_* = g_* f_*$ $\quad (g : Y, B \to Z, C)$.

3. $\partial f_* = (f \mid A)_* \partial \quad (\partial : H_p(X, A) \to H_{p-1}(A))$.

4. *Exactness*. The sequence

$$\ldots \to H_p(A) \to H_p(X) \to H_p(X, A) \xrightarrow{\partial} H_{p-1}(A) \to \ldots$$

is exact.

5. *Homotopy*. If f and g are homotopic then $f_* = g_*$.

6. *Excision*. If U is open and $\overline{U} \subset int\, A$ then inclusion induces an isomorphism between $H_p(X - U, A - U)$ and $H_p(X, A)$.

7. *Dimension*. If P is a single point then

$$H_p(P) = \begin{cases} 0 \text{ if } p > 0 \\ \text{the coefficient group if } p = 0. \end{cases}$$

6. A homology theory without the dimension axiom is called a generalised homology theory, e.g. bordism.

7. There is an ambiguous sign here, but see Section 1.6.

8. For example, $\varepsilon_a(aba^{-1}b^{-2}) = 0$, $\varepsilon_b(aba^{-1}b^{-2}) = -1$.

9. More precisely, $C_n(K, L) = C_n(K)/i\,C_n(L)$, where $i = C_n(L) \to C_n(K)$ is induced by inclusion.

10. Notice that this is a finite sum because of the conditions imposed upon the definition of a cell complex.

11. Chains c such that $\partial c = 0$ are called *cycles* and are often designated by z. A chain $c = \partial c'$ is called a *boundary*. Since $\partial^2 = 0$ a boundary may quite properly be called a boundary cycle.

12. For a proof see Eilenberg and Steenrod. Note that the isomorphism is not natural.

13. Beware, d is not a cellular map!

14. This formula is associative, but only anti-commutative up to a coboundary.

15

16. No manifold has been discovered which has failed to be triangulable.

17. Recall that the *link* of a simplex $a_0 a_1 \ldots a_p$ is the collection of simplexes $a_{p+1} \ldots a_q$ such that $a_0 \ldots a_p a_{p+1} \ldots a_q$ lies in K. Look at an example of the link of a *1-simplex* in a *3-manifold* and you will see where the term 'link' comes from. Usually it is assumed that the combinatorial structure of the link sphere is standard, but this is unimportant since only the homology of the link and star are needed. For details, see Spanier.

18. The symbol $\wedge$ is a magic helmet or 'tarnhelm' which makes the wearer invisible.

19. To see this, put $x = 1$, $x \neq a$, in $\omega(r)$. The resulting word collapses to 1 by cancelling pairs of the form aa^{-1} or $a^{-1}a$. Let this define the pairing and the paths $\alpha, \beta, \ldots$. Clearly these paths may be chosen disjoint.

20. ε_a and ε_b are particular forms of the more general $\varepsilon_{i_1 i_2 \ldots i_k}$ considered in the chapter on Massey products.

21. The cell r is oriented so that 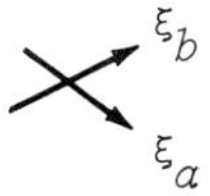 is a positive intersection.

2: AN INTRODUCTION TO HOMOTOPY

'A central problem of topology is the determination of the homotopy classes of mappings'

Solomon Lefschetz *Introduction to Topology*

In Sections 1 and 2 of this chapter we define the homotopy groups and describe, without too much proof, some of their well-known properties.

Definition 2.1.1 **The homotopy groups**

If X, Y are spaces let $[X; Y]$ denote the set of homotopy classes of maps $f : X \to Y$. This notation can be extended to include sets of spaces. So, for example $[I, 0, 1; X, x_0, x_0]$ is the set of homotopy classes of loops in X based at x_0.

If $f : X \to Y$ is a continuous map let $[f]$ denote its class in $[X; Y]$.

In terms of this notation the **absolute homotopy groups** [1] are given by

$$\pi_n(X, x_0) = [S^n, *; X, x_0].$$

The base point $*$ can be chosen arbitrarily but for definiteness let $* = (1, 0, \ldots, 0)$.

The set π_n is a group because of the map $p : S^n \to S^n \vee S^n$ which collapses the equator $x_{n+1} = 0$ to the base point.

If $\alpha = [f]$ and $\beta = [g]$ lie in $\pi_n(X, x_0)$ let the product $\alpha\beta = [h]$ where h is the composition $S^n \xrightarrow{p} S^n \vee S^n \longrightarrow X$, the right-hand map being α on the first factor and β on the second factor.

The product is associative for all n and Abelian if $n > 1$. There is an identity, the class of the trivial map $\iota : S^n \to X$, $\iota(S^n) = x_0$, and each element has an inverse. The inverse to the element $[f]$ is given by $[f \circ T]$ where T is an orientation reversing homeomorphism of S^n.

These results and others are easier to prove if the following alternative definition in terms of cubes is used. Consider I^{n-1} as the subset of I^n for which $x_n = 0$. Then

$$\pi_n(X, x_0) = [I^n, \partial I^n; X, x_0]$$

A cube I^n is split into two subcuboids by the hyperplane $x_1 = \frac{1}{2}$. This allows the definition of the product to be given by the following rule:

If $f, g : I^n, \partial I^n \longrightarrow X, x_0$ let

$$f * g\,(x_1, x_2, \ldots, x_n) = \begin{cases} f\,(2x_1, x_2, \ldots, x_n) & x_1 \leq \frac{1}{2} \\ g\,(2x_1 - 1, x_2, \ldots, x_n) & x_1 \geq \frac{1}{2}. \end{cases}$$

Then $[f]\,[g] = [f * g]$.

The cube definition also facilitates the definition of the relative homotopy groups.

Let J^{n-1} be the subset of ∂I^n consisting of all points with either $x_n > 0$ or $x_n = 0$ and $x_i = 0$ or 1 for some $i \neq n$. So J^{n-1} is the closure of the complement of I^{n-1} in ∂I^n see Figure 1.

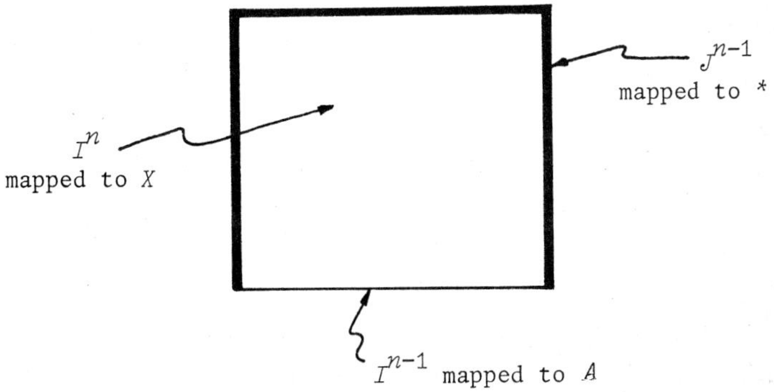

Figure 1.

Then if $A \subset X$ and $x_0 \in A$ let

$$\pi_n(X, A, x_0) = [I^n, I^{n-1}, J^{n-1}; X, A, x_0].$$

The same definition of product makes $\pi_n(X, A, x_0)$ into a group if $n \geq 2$ which is Abelian if $n \geq 3$.

If the base point x_0 lies in the same path component of X as the base point y_0, then $\pi_n(X, x_0)$ and $\pi_n(X, y_0)$ are isomorphic. Consequently, unless special emphasis is needed we shall drop the base point and write $\pi_n(X)$ for $\pi_n(X, x_0)$. Similarly, if x_0 and y_0 lie in the same path component of A then $\pi_n(X, A, x_0)$ is isomorphic to $\pi_n(X, A, y_0)$.

2.1.2 The exact sequence of a pair

The correspondence $X, A \longrightarrow \pi_n(X, A)$ is a functor in the usual sense from spaces to groups and like homology there is also an exact sequence.

Let $\partial : \pi_n(X, A) \longrightarrow \pi_{n-1}(A)$ be the map obtained by restricting a representative map $f : I^n, I^{n-1}, J^{n-1} \longrightarrow X, A, x_0$ to I^{n-1}. If $\pi_n(A) \longrightarrow \pi_n(X)$ and $\pi_n(X) \longrightarrow \pi_n(X, A)$ are induced by inclusion then there is an exact sequence

$$\ldots \longrightarrow \pi_n(A) \longrightarrow \pi_n(X) \longrightarrow \pi_n(X, A) \xrightarrow{\partial} \pi_{n-1}(A) \longrightarrow \ldots \quad n \geq 2.$$

We shall have more to say on the tail end of this sequence later.

2.2 Relations with homology

The homotopy groups π_n are related to the homology groups H_n by the Hurewicz map, $h : \pi_n(X) \longrightarrow H_n(X)$ which is defined as follows: represent an element of $\pi_n(X)$ by a map $f : S^n \longrightarrow X$. Choose the positive generator t of $H_n(S^n)$, the infinite cyclic group. Then $h[f]$ is $f_*(t)$.

Theorem 2.2.1 (The Hurewicz isomorphism)

Let K *be a path connected complex.*

1. If $n > 1$ *and* $\pi_i(K) = 0$ *for* $i < n$ *then the Hurewicz map* $\pi_n(K) \longrightarrow H_n(K)$ *is an isomorphism.*

2. If $n = 1$ *the Hurewicz map is onto and has kernel the commutator subgroup* $[\pi, \pi]$. *So* $H_1 = \pi/[\pi,\pi]$, [2].

The π_n are stronger than the H_n because of Whitehead's theorem:

Theorem 2.2.2 (Whitehead)

If $f : K \longrightarrow L$ *is a continuous map between connected complexes and induces isomorphisms* $f_* : \pi_i(K) \longrightarrow \pi_i(L)$ *for* $i = 1,2,\ldots,n$ *where* $n \geq \max(\dim K, \dim L)$ *then* f *is a homotopy equivalence.*

This statement is not true if f induces a homology isomorphism, although it is true for special cases, e.g. where π_1 is trivial or abelian.

Of course a price must be paid for this power. In general the π_n are very hard to calculate. However, π_1 and π_2 are in principle easy to calculate and mathematically very interesting. We shall spend the rest of the chapter looking at these two groups.

2.3 Pictures

In this section we consider representations of homotopy group elements in terms of certain planar graphs called **pictures**.

If K is any compact connected cell complex then K has the homotopy type of a cell complex with just one 0-cell which we will always denote by e and will always take as the base point of K. Moreover $\pi_i(K, e)$ is isomorphic to $\pi_i(K^{(i+1)}, e)$ so in order to study the fundamental group it is only necessary to look at 2-complexes with one 0-cell.

Using the notation developed in the previous chapter let $K = \{A \mid R\}$ be a 2-complex with base vertex e. Let

$g : S^1, 1 \longrightarrow K^1, e$ be a map. Then as usual we can associate with g a word $w(g)$ in the variables $A \cup A^{-1}$. Specifically, assume that K and S^1 are triangulated so that g is simplicial. If $\hat{a}$ lies in the interior of a 1-simplex lying in A, then $w(g)$ is determined by the inverse image $g^{-1}\bigsqcup_{a \in A} \hat{a}$ which is a finite set.

This is the first example of a concept which we shall meet again in this and later chapters. The map g is said to be **transverse** to the cell a at the point $\hat{a}$.

The word $w(g)$ determines the class $[g]$ of g in $\pi(K^1)$ and conversely up to the introduction or deletion of cancelling pairs aa^{-1} or $a^{-1}a$.

Now suppose that g is homotopic to e in K. Then g extends to a map $f : D^2, S^1 \longrightarrow K, K^1$. We shall now alter f to make it as agreeable as possible.

Firstly pick a point $\hat{r}$ in the interior of each 2-cell $r \in R$. By a homotopy keeping $f \mid S^1 = g$ fixed assume that f is transverse to the cells r at $\hat{r}$. This means that if D_r is a small enough disc containing $\hat{r}$ then $f^{-1}\bigsqcup_{r \in R} D_r$ is a finite collection $\{N\}$ of disjoint discs in the interior of D^2 and for each N $f \mid N$ is a homeomorphism from N onto some D_r.

Now by a radial expansion assume that $f \mid N$ satisfies:

1. $f(N) = r$,
2. $f \mid \mathit{int}\, N$ is a homeomorphism from $\mathit{int}\, N$ to $\mathit{int}\, r$.

and 3. $f \mid \partial N$ is transverse to K^1 as considered above.

This means that $w(f \mid \partial N)$ is equal to some cyclic permutation of the word $w(r)$ or $w(r)^{-1}$.

Let $X = D^2 - \bigsqcup \mathit{int}\, N$, then X is a surface and f maps X into the space $Y = K - \bigsqcup_{r \in R} \hat{r}$.

But K^1 is a deformation retract of Y. So by a homotopy keeping $f \mid \overline{D^2 - X}$ fixed assume that $f(X) \subset K^1$ and that $f \mid X$

is transverse to each a at $\hat{a}$. This means as before that $\hat{a}$ is the centre of some 1-simplex into which f is simplicial. Then $f^{-1}\bigcup_{a \in A} \hat{a}$ is a 1-manifold. Call such a map **transverse to K**.

Altogether the discs N labelled by the $r \in R$ and the arcs $f^{-1}\bigcup_{a \in A} \hat{a}$ labelled by the $a \in A$ form a graph (with fat vertices) in D^2 called a **disc picture**.

The arcs $f^{-1}\bigcup_{a \in A} \hat{a}$ are oriented transversely but it is more convenient when drawing the picture to orient the arcs along their length. The convention adopted is the **left-hand rule**. The transverse orientation points to the left of a swimmer swimming in the positive direction along the arc.

Summing up the above discussion we see that every $f : D^2 \longrightarrow K$ is homotopic modulo its boundary to a transverse map and that associated with every transverse map is a disc picture consisting of vertices oriented + or - and labelled by the $r \in R$ and oriented edges labelled by the $a \in A$. The form of the graph near each vertex is determined as follows:

2.3.1 Example (the relation spider)

Let $\phi : D^2 \longrightarrow K$ be the characteristic map of the cell r. Then $w(\phi \mid S^1)$ spells out the word $w(r)$ or a cyclic permutation.

There is no ambiguity if every cell r comes equipped with a **tag** attached to the base point e. The word $w(r)$ is then read off in an anticlockwise fashion starting after the tag. So, for example, if $w(r) = aba^{-1}b^{-1}b^{-1}$ then the disc picture of ϕ is pictured in Figure 2.

The oppositely oriented disc picture with word $bbab^{-1}a^{-1}$ is obtained by reflexion in a line and is illustrated on the right of Figure 2.

In general, a disc picture will consist of a number of these spiders corresponding to various 2-cells. The feet of the spiders will be attached coherently by arcs to other feet or to the

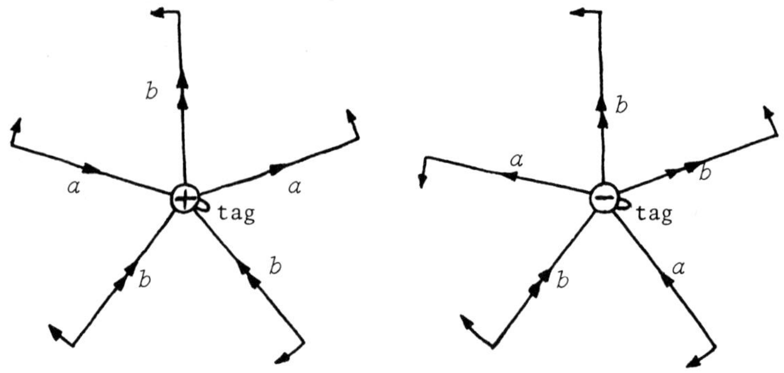

Figure 2. A spider and an antispider.

boundary of the disc. Also, there could be arcs or loops labelled by the $a \in A$ but not associated with any 2-cell.

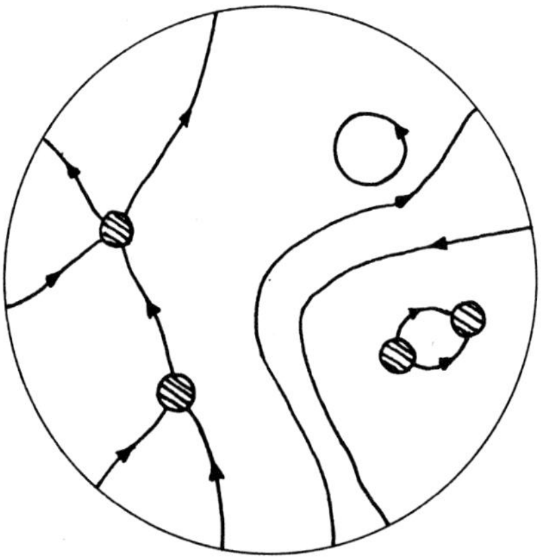

Figure 3. A general disc picture.

Definition 2.3.2

A disc picture which is disjoint from the boundary of the disc and hence spells out the empty word is called a spherical picture.

2.3.3 Examples of spherical pictures

1. A spider and its mirror image form a spherical picture called a **cancelling pair** as in Figure 4, where $w(r) = abab^{-2}$.

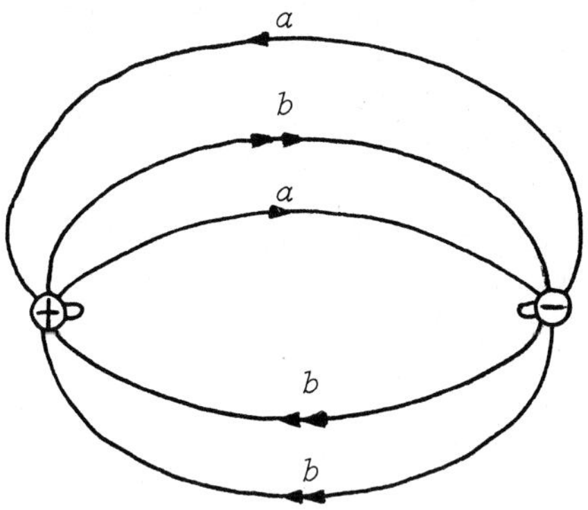

Figure 4. A cancelling pair of spiders.

2. If $w(r)$ is a proper power, say r^k, then a spider and its mirror image can form a spherical picture in k ways. Only one of these is called a cancelling pair and that is the one where the base point tags are opposite each other.

For instance, associated with the relation $a^2 = 1$ are two spherical pictures as in Figure 5. We shall see later that these are essentially different.

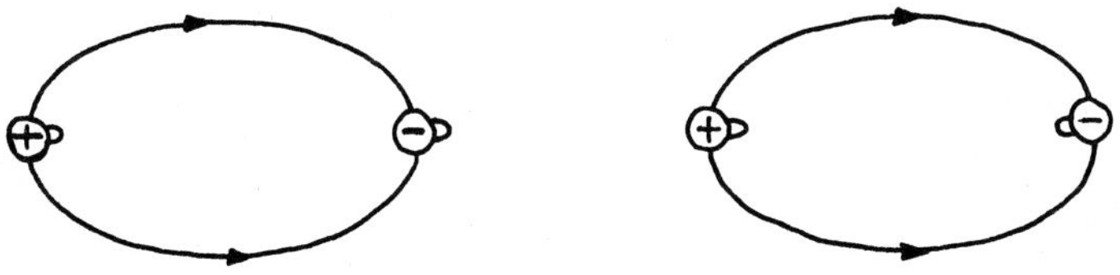

Figure 5. Spherical pictures associated with $a^2 = 1$. The one on the right is a cancelling pair.

2.3.4 Realisation of pictures

Given a transverse map of the disc into the complex K we have seen how to associate a picture with the map. Conversely, given a picture there is a way of associating with it a transverse map. The proof proceeds inductively on the measure of complexity n of the picture. Here n is the total number of 2-cell discs and isolated 1-cell arcs. If $n = 1$ there is no difficulty. Otherwise split the disc into two discs with lower complexity by an arc which avoids the 2-cell discs and cuts the 1-cell arcs transversely. Now, using induction, define the maps on the sub-discs and glue together via their common portion of boundary.

2.3.5 Examples

1. Let $K = \{a, b \mid a^2, b^2, (ab)^2\}$. The disc diagram in Figure 6 illustrates the fact that $ab = ba$ in $\pi(K)$.

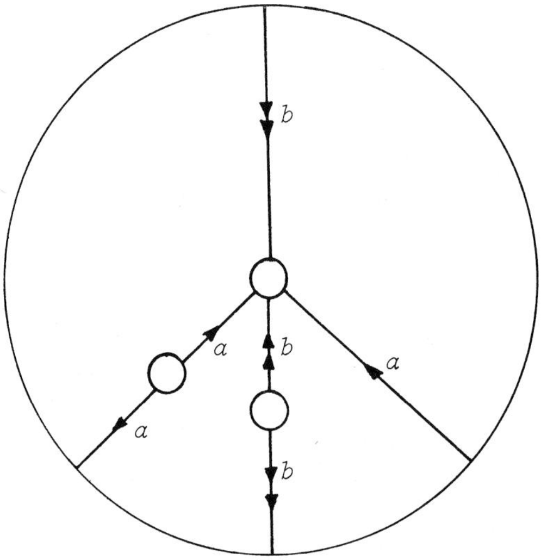

Figure 6. $ab = ba$.

2. The complex

$$K = \{ a,b,c \mid r = [a,b] b^{-1}, s = [b,c] c^{-1}, t = [c,a] a^{-1} \}$$

is an unlikely presentation of the trivial group [3] . The picture illustrated in Figure 7 shows that $[a]$ is trivial in $\pi_1(K)$.

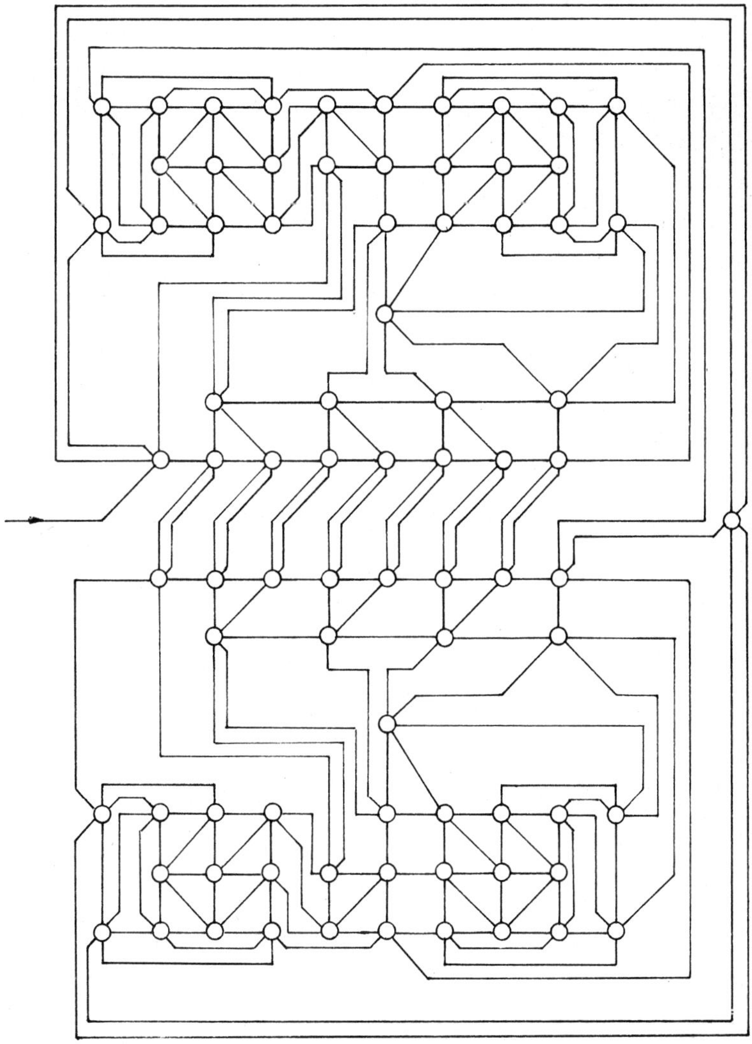

Figure 7.

2.4 Pictures and homotopies

We have seen that associated with any transverse map is a picture and conversely. Now we answer the following question: if two transverse maps are homotopic, how are their pictures related?

We firstly describe three alterations or **moves** which can be applied to pictures and which do not change the homotopy class of the corresponding maps $D^2, S^1 \longrightarrow K^2, K^1$. Then we show that if two such maps are homotopic their pictures can be changed by a sequence of these moves.

M1. The birth and death of a pair of cancelling 2-cell discs.

This move introduces or deletes a pair of oppositely oriented spiders with corresponding feet locked in embrace.

If the spiders correspond to relations which are proper powers, then care must be taken to ensure that the tags are opposite one another.

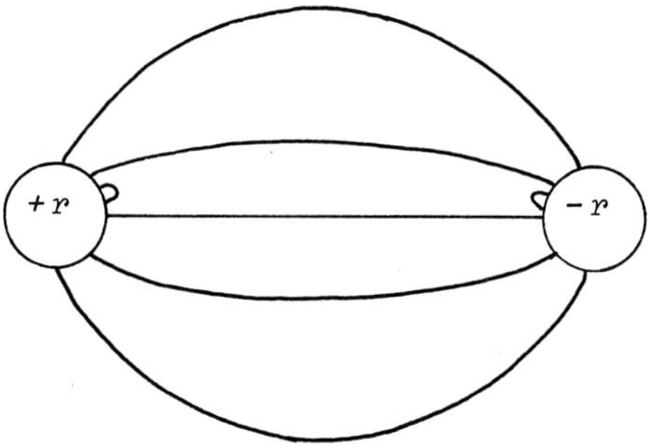

Figure 8. A newly-born cancelling pair of 2-cell discs.

M2. The birth or death of a 1-cell loop.

This move introduces or deletes a small, isolated 1-cell loop.

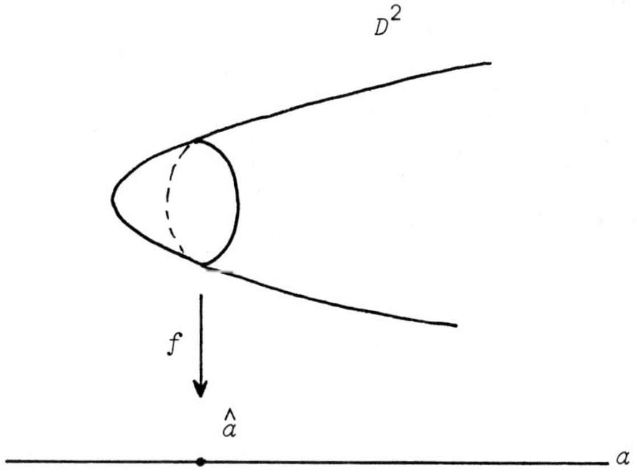

Figure 9.

M3. The bridge move.

This move is illustrated in Figure 10 and can be described as follows: Let $ABCD$ be a rectangle continuously embedded in D^2 which only meets the picture in the sides AB, CD which are oppositely oriented portions of 1-cell arcs both labelled with the

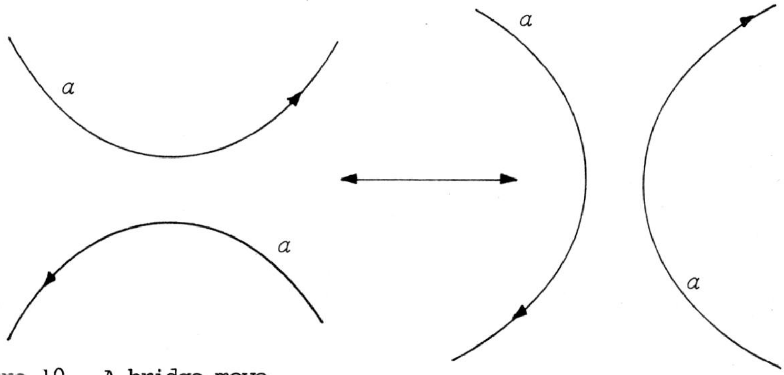

Figure 10. A bridge move.

same 1-cell a. Alter the *a-arcs* by removing AB, CD and replacing with AD, CB.

Now consider a homotopy keeping S^1 fixed between two transverse maps $f, g : D^2, S^1 \longrightarrow K^2, K^1$. This may be represented by a map $h : D^2 \times I, \partial D^2 \times I \longrightarrow K^2, K^1$ such that for all x in D, $h(x, 0)$ is $f(x)$ and $h(x, 1)$ is $g(x)$. Now apply the same procedure to h which was applied to maps on the disc. Assume, by a homotopy keeping $h \mid \partial(D^2 \times I)$ fixed, that h is transverse to K. This means that for a small disc D_r about $\hat{r}$, $h^{-1}(D_r)$ is a collection of tubes in $D^2 \times I$. Some of these tubes may be joined up to form tori, others may join up 2-cell discs in the top and bottom pictures at $D^2 \times 0$ and $D^2 \times 1$.

Also the inverse images $f^{-1}\hat{a}$, where a is a 1-cell, will be surfaces. These may be closed surfaces or they may have boundary. The boundary will consist of 1-cell arcs in $D^2 \times 0$ and $D^2 \times 1$ and arcs running the length of the 2-cell tubes. This means that a cross-section of a 2-cell tube will consist of a relation spider, the legs of the spider forming fins along the tubes.

Call this three-dimensional collection a **homotopy picture** joining the f-picture downstairs to the g-picture upstairs. To any transverse homotopy corresponds a homotopy picture and conversely to every homotopy picture corresponds a transverse homotopy.

Now a homotopy picture could be an incredibly complicated collection of knotted and linked tubes and surfaces, but if the 2-cell tubes and the 1-cell surfaces are repositioned so that the height function is Morse and if horizontal cross-sections $D^2 \times \{t\}$ $0 \leq t \leq 1$ are moved upward, then the resulting pictures will change only at a finite number of places in one of the three ways illustrated in Figures 11, 12 and 13.

To sum up, a picture can represent three things:

1. A homotopy to zero of a map $f : S^1 \longrightarrow K^1 \subset K^2$.
2. An element of the relative homotopy group $\pi_2(K^2, K^1)$.
3. An element of the absolute homotopy group $\pi_2(K^2)$ if the picture is spherical.

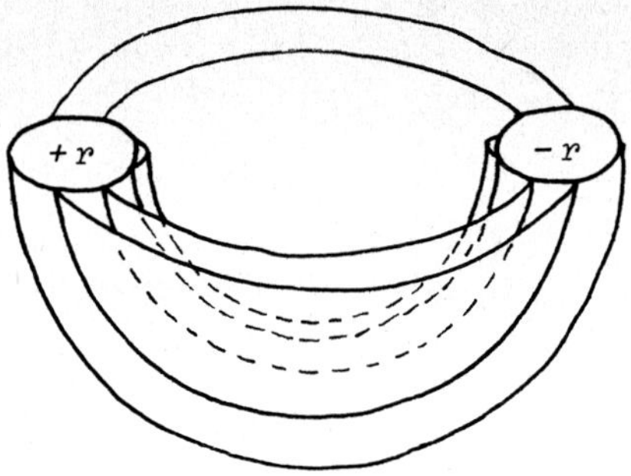

Figure 11. Birth of a cancelling pair of 2-cell discs.

Figure 12. Death and birth of isolated 1-cell arcs.

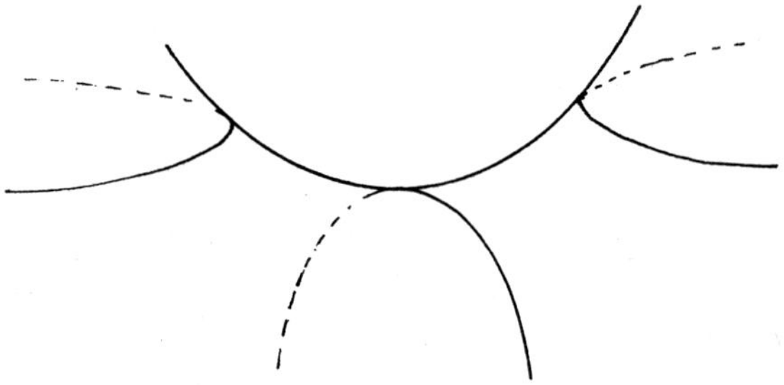

Figure 13. A bridge move.

Two pictures represent homotopic maps if and only if one can be obtained from the other by a sequence of moves M1, M2 and M3 .

2.4.1 Examples

1. Let K be the complex $\{a \mid r = 1, s = a\}$. So K is the wedge of a disc s with a sphere r. The boundary of the disc is the loop a. If the interior of the disc is removed to obtain the

subcomplex $L = \{a \mid r = 1\}$ then the following spherical picture represents a non-trivial element x of $\pi_2(L)$.

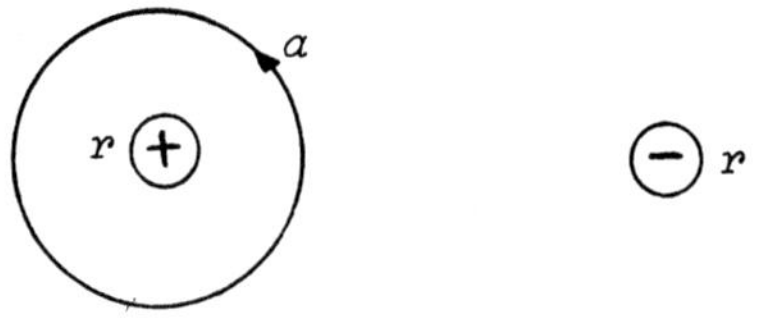

Figure 14. A non-trivial element x of $\pi_2(L)$. The 2-cell discs corresponding to the relation $1 = 1$ are oppositely oriented. Note that x is homologous to zero under the Hurewicz map.

Under the inclusion map $\pi_2(L) \longrightarrow \pi_2(K)$ the element x goes to zero. We can see this by the following sequence of *M moves* :

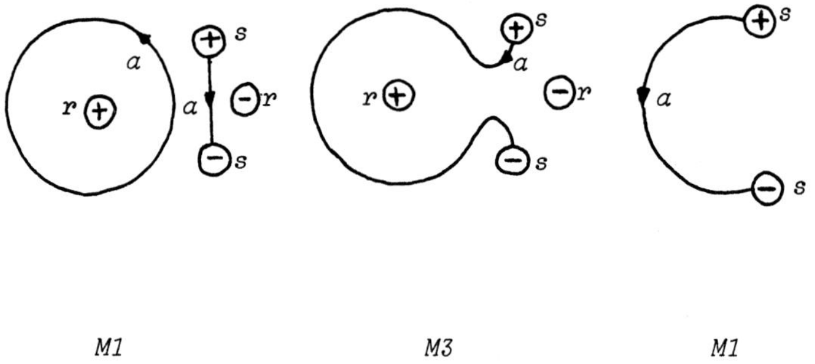

Another *M1* completes the homotopy.

This example illustrates the perhaps rather surprising fact that adding a 2-cell can kill elements of π_2, [4].

2. Let $K = \{a \mid a^n = 1\}$ then in later chapters we shall show that $\pi_2(K)$ is free abelian of rank $n-1$. The generators are

represented by two opposite spiders their feet joined up as if to represent a cancelling pair but with one tag rotated through an angle of $2\pi i/n$, $i = 1, 2, \ldots, n-1$.

2.4.2 Boundary moves

We conclude this section by looking at what happens if the boundary of a map $f : D^2, S^1 \longrightarrow K^2, K^1$ is allowed to move during a homotopy.

Suppose the boundary word contains a cancelling pair $\ldots aa^{-1} \ldots$. Then the associated picture will have two a arcs in opposite directions next to one another, meeting the boundary. By a bridge move if necessary, it can be assumed that the picture looks like Figure 15 at this part of the boundary, where the shaded part contains no other portion of the picture.

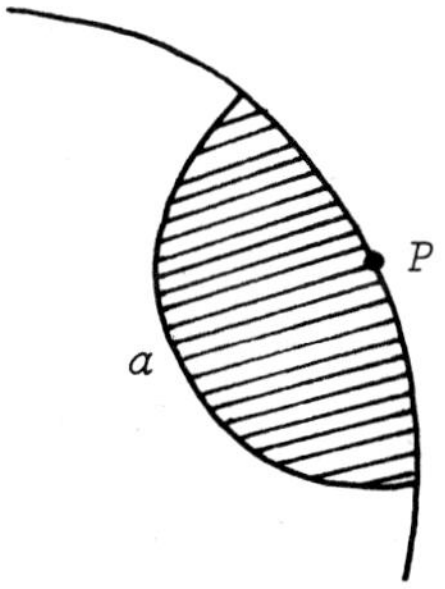

Figure 15.

This disc can now be removed or inserted, corresponding to the deletion or insertion of the pair aa^{-1} in the boundary word. This procedure cannot be used if P is the base point 1 on S^1 . Then the boundary word is a conjugate aga^{-1} .

2.5 Path Transportation

In this section we return to the remark made earlier, that for path connected spaces homotopy groups based on different base points are isomorphic. This isomorphism is given by transporting elements of the homotopy group along a path joining the two points.

Write S^n as the union of a **western hemisphere**

$$W = \{(x_1, x_2, \ldots, x_n) \in S^n \mid x_1 \leq 0\}$$

and an **eastern hemisphere**

$$E = \{(x_1, x_2, \ldots, x_n) \in S^n \mid x_1 \geq 0\}.$$

Let two points $x = (x_1, \ldots, x_n)$, $y = (y_1, \ldots, y_n)$ in S^n be equivalent if and only if the following conditions are satisfied:

If $x, y \in W - E$ then $x = y$;

if $x, y \in E$ then $x_1 = y_1$;

if $x \in W - E$, $y \in E - W$ or $x \in E - W$ and $y \in W - E$,

then x and y cannot be equivalent.

Then the quotient space X is the wedge of a sphere with an arc, $0 \leq x_1 \leq 1$. Let $p : S^n \longrightarrow X$ be the identification map.

Let e_0 and e_1 be two points in the complex K and let $\alpha : I \longrightarrow K$ be a path from e_0 to e_1.

If $f : S^n, 1 \longrightarrow K, e_1$ is a map let $\alpha_{\#} f : S^n, 1 \longrightarrow K, e_0$ be the map which is the composition of p with the map which is f on the sphere and $t \longrightarrow \alpha(-t)$ on the arc of X.

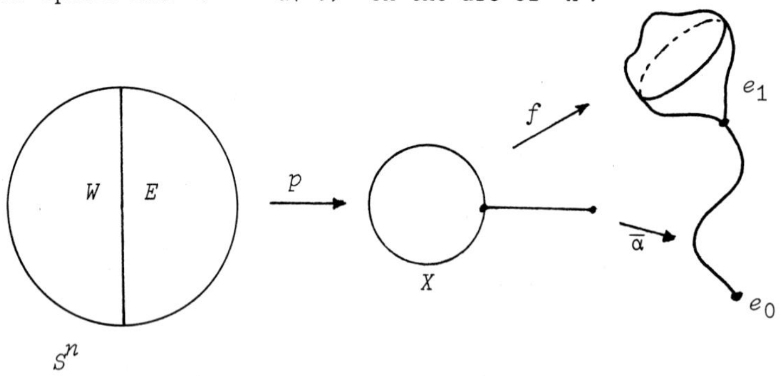

Figure 16. Transportation $\alpha_{\#} f$

If α, β are joinable paths in K, $\alpha(1) = \beta(0)$ then it is not hard to see that

$$(\alpha\beta)_{\#} f \simeq \alpha_{\#} \beta_{\#} f .$$

Moreover, if $\alpha_1 \simeq \alpha_2$ then $\alpha_{1\#} f \simeq \alpha_{2\#} f$. So if α is a loop in K based at e then we have an action of the fundamental group on $\pi_n(K, e)$ given by $[\alpha] \cdot [f] = [\alpha_{\#} f]$.

If $n = 1$ this action is conjugation, i.e.

$$g \cdot h = g h g^{-1} .$$

If $n = 2$ then the action can be visualised with pictures as follows: If p_f is a spherical picture representing the transverse map $f : D^2, S^1 \longrightarrow K, e$ and a is a 1-cell then

$[a] \cdot [f]$ is represented by 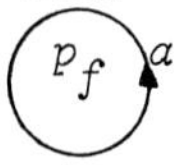

and $[a]^{-1} \cdot [f]$ is represented by 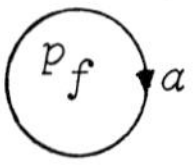

More generally, an element of $\pi_1(K, e)$ can be represented as a word $[a_1 a_2 \ldots a_k]$ $a_i \in A \cup A^{-1}$ and the action can be pictured as repeated applications of the above.

2.5.1 The action of $\pi_1(K^1)$ on $\pi_2(K, K^1)$

If $[f]$ is an element of $\pi_2(K, K^1)$ and a is a 1-cell in K^1 representing some element $[a]$ in $\pi_1(K^1)$ then $[a] \cdot [f] = [a_{\#} f]$ as before, [5].

Pictorially, this action can be described by Figure 17. Once again the general action is defined by repetition.

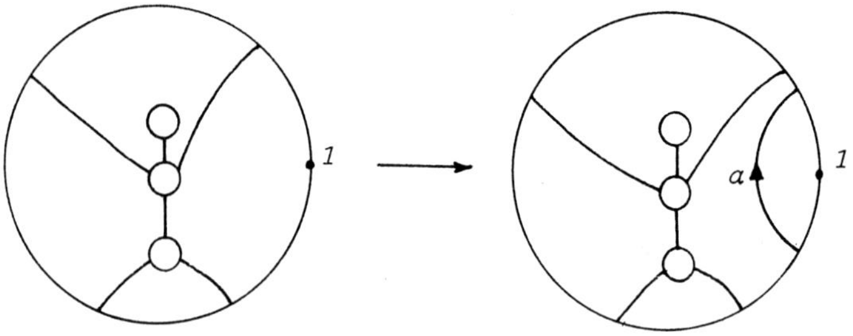

Figure 17. The action of $\pi_1(K^1)$ on $\pi_2(K, K^1)$.

2.6 Crossed modules

In this section we look at the algebraic conditions satisfied by the relative group $\pi_2(K, K^1)$ and call the general object satisfying these conditions a crossed module.

Let us adopt the following notation. K as usual denotes a 2-dimensional cell complex with one 0-cell e. Let $\pi_2 = \pi_2(K)$ and $C = \pi_2(K, K^1)$. Let $F = F(A) = \pi_1(K^1)$ and let $\pi = \pi_1(K)$. Then the exact sequence of the pair K, K^1 yields:

$$0 \longrightarrow \pi_2 \xrightarrow{i} C \xrightarrow{\partial} F \xrightarrow{p} \pi \longrightarrow 1 .$$

The inclusion map i embeds π_2 as a normal abelian subgroup of C and $\partial(c)$ is the class of the boundary word of c in C. If $N = C/\pi_2$ then there is a free resolution of π_1 corresponding to the presentation K,

$$1 \longrightarrow N \longrightarrow F \xrightarrow{p} \pi \longrightarrow 1 .$$

The subgroup N is the usual closure in F of the relations $w(r)$. A typical element of N is a product

$$\Pi g_i \, w(r_i)^{\pm 1} \, g_i^{-1}, \quad g_i \in F, \quad r_i \in R .$$

The homomorphism $\partial : C \longrightarrow F$ and the action of F on C interact as follows:

$$\partial(f \cdot c) = f(\partial c)f^{-1} \quad \ldots \quad 1.$$

$$\partial c \cdot b = cbc^{-1} \quad \ldots \quad 2.$$

Definition 2.6.1 Crossed F-modules [6]

A **crossed F-module** is a group C together with an action of F on C and a homomorphism $\partial : C \longrightarrow F$ satisfying 1. and 2. above.

We talk of C as the crossed module although strictly speaking we should talk of C, the map ∂ and the action of F as the crossed module.

This notion was introduced by Whitehead in 1941.

2.6.2 Examples of crossed modules

Apart from the canonical example of a crossed module as the 2-dimensional relative homotopy group, the following two examples show that this concept generalises both the notion of a normal subgroup and the notion of a (an uncrossed) *F-module*.

1. Let $C = N$, be a normal subgroup of $F = G$. Then C is a crossed *G-module* if $\partial : N \longrightarrow G$ is taken to be inclusion and the action of G on N is by conjugation, i.e. $g \cdot n = gng^{-1}$.

2. If C is an *F-module* in the usual sense then the constant map $c \to 1$, $c \in C$ makes C into a crossed *F-module*.

Definition 2.6.3 Morphisms of crossed modules.

A **morphism** from the crossed *F-module* C to the crossed *F'-module* C' is a pair of homomorphisms $(\phi, \psi) : (C, F) \to (C', F')$ which makes the following square commute:

$$\begin{array}{ccc} C & \xrightarrow{\ \phi\ } & C' \\ {\scriptstyle\partial}\big\downarrow & & \big\downarrow{\scriptstyle\partial} \\ F & \xrightarrow{\ \psi\ } & F' \end{array}$$

and such that $\phi(f \cdot c) = \psi(f) \cdot \phi(c)$, for all f in F and all c in C.

Definition 2.6.4 **Free crossed modules**

A crossed *F-module* C is said to be **free** on the subset R of C if given an *F-module* D and a function $\alpha : R \longrightarrow D$ there is a unique homomorphism $\phi : C \longrightarrow D$ extending the function α such that the pair $(\phi, 1) : (C, F) \longrightarrow (D, F)$ is a morphism of crossed *F-modules*.

The subset R is called a **basis** for C.

The following theorem says that free crossed modules exist and are unique.

Theorem 2.6.5 *Given any set* R, *any group* F *and any function* $w : R \rightarrow F$, *there is a unique free crossed F-module* C *with basis* R *such that* $\partial|R = w$.

Proof. The proof of 2.6.5 motivates the introduction of several new concepts such as Peiffer elements, relation identities, etc.

Let H be the free operator group on R with F acting on the left. Then H is the free group on the set $Y = \{ f \cdot r \mid f \in F, r \in R \}$ and the F operation is defined by the rules:

$$f \cdot (hk) = (f \cdot h)(f \cdot k) \qquad h, k \in H, \quad f \in F$$

$$f \cdot (g \cdot h) = (fg) \cdot h \qquad f, g \in F, \quad h \in H.$$

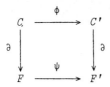

It is also convenient to write

$$(f \cdot r)^{-1} = f \cdot r^{-1} \quad f \in F, \quad r \in R \quad \text{and} \quad 1 \cdot r = r.$$

Notice that both Y and Y^{-1} are invariant under F.

A typical element of H, called a **relation string** for reasons which will be apparent later, is a word $(f_1 \cdot r_1)(f_2 \cdot r_2) \ldots (f_n \cdot r_n)$ where $f_i \in F$ and $r_i \in R \cup R^{-1}$.

Definition 2.6.6 Identities and Peiffer identities

There is a homomorphism $\theta : H \to F$ given on the generators by $\theta(f \cdot r) = f\omega(r) f^{-1}$.

The elements of $E = ker\ \theta$ are called **relation identities,** or **identities** for short.

The **Peiffer identities** are the elements of E of the form $p = hkh^{-1}(\theta(h) \cdot k^{-1})$ [7].

If $h, k \in Y \cup Y^{-1}$ then p is called a **basic Peiffer identity.**

Lemma 2.6.7 *Let P be the subgroup of H generated by the Peiffer identities. Then P is*

1. *the normal closure in H of the basic Peiffer identities;*
2. *invariant under the action of F on H.*

Proof

1. Firstly, let us show that P is normal in H. If $p = hkh^{-1}\theta(h) \cdot k^{-1}$ is a Peiffer identity and $u \in H$ then

$$upu^{-1} = \left\{(uh)k(uh)^{-1}(\theta(uh) \cdot k^{-1})\right\}\left\{u(\theta(h) \cdot k)u^{-1}\theta(u) \cdot (\theta(h) \cdot k^{-1})\right\}^{-1}.$$

The expressions inside the braces are both Peiffer, so $upu^{-1} \in P$.

Hence if $\tilde{P}$ is the normal closure in H of the basic Peiffer identities then $\tilde{P} \subset P$ and $\tilde{P}$ is normal in P.

Work now modulo $\tilde{P}$ which implies that $ab \equiv (\theta(a) \cdot b)a$ or $\theta(a) \cdot b \equiv aba^{-1}$ if $a, b \in Y \cup Y^{-1}$. Let $p = hkh^{-1}\theta(h) \cdot k^{-1}$.

be an arbitrary Peiffer identity. Write $k = b_1 \dots b_\ell$ where $b_i \in Y \cup Y^{-1}$. Then $p = s_1 s_2 \dots s_\ell t_\ell t_{\ell-1} \dots t_1$ where $s_i = h b_i h^{-1}$ and $t_i = \theta(h) \cdot b_i^{-1}$, $i = 1, \dots, \ell$. Writing $h = a_1 \dots a_m$ where $a_i \in Y \cup Y^{-1}$ then

$$\begin{aligned} s_\ell t_\ell &= a_1 \dots a_m b_\ell a_m^{-1} \dots a_1^{-1} \theta(a_1 \dots a_m) \cdot b_\ell^{-1} \\ &\equiv a_1 \dots a_{m-1} c_m a_{m-1}^{-1} \dots a_1^{-1} \theta(a_1 \dots a_{m-1}) \cdot c_m^{-1} \end{aligned}$$

mod $\tilde{P}$ where $c_m = \theta(a_m) \cdot b_\ell \in Y \cup Y^{-1}$.

So by induction on m $s_\ell t_\ell \equiv 1$ and by induction on ℓ, $p \equiv 1$. Hence $\tilde{P} = P$.

2. If $p = h k h^{-1} \theta(h) \cdot k^{-1}$ is Peiffer and $f \in F$ then

$$\begin{aligned} f \cdot p &= (f \cdot h)(f \cdot k)(f \cdot h)^{-1} f \cdot (\theta(h) \cdot k^{-1}) \\ &= (f \cdot h)(f \cdot k)(f \cdot h)^{-1} \theta(f \cdot h) \cdot (f \cdot k)^{-1}, \end{aligned}$$

which is also Peiffer and so P is invariant under F.

Proof of 2.6.5 (concluded)

We are now in a position to complete the construction of the free crossed *F-module* on the set R promised by 2.6.5.

Let $C = H/P$. An element of C is therefore a class $[y_1 y_2 \dots y_k]$ where $y_i \in Y \cup Y^{-1}$ and two strings are in the same class if and only if one can be obtained from the other by a sequence of Peiffer moves. That is, an interchange of the form $y_i y_{i+1} \longleftrightarrow (\theta(y_i) \cdot y_{i+1}) y_i$

or $\quad y_i y_{i+1} \longleftrightarrow y_{i+1} (\theta(y_{i+1}^{-1}) \cdot y_i)$,

together with the usual cancelling or insertion of pairs $y_i y_{i+1}$ if $y_{i+1} = y_i^{-1}$.

Then C has an *F-operation* induced by the *F-operation* on H. Moreover, there is a map $\partial : C \longrightarrow F$ given by

$$\partial [y_1 \dots y_k] = \theta(y_1) \dots \theta(y_k).$$

Dividing out by the Peiffer identities makes H into a crossed *F-module* since if $h \in H$

$$\partial(f \cdot [h]) = f[\theta(h)]f^{-1} = f(\partial[h])f^{-1} \quad \text{and}$$

$$\partial[h] \cdot [k] = \theta(h) \cdot [k] = [\theta(h) \cdot k] = [hkh^{-1}] = [h][k][h]^{-1}.$$

Furthermore, it is a simple matter to check that C is a free crossed module on R and uniqueness follows from the usual universality argument.

Theorem 2.6.8 (Whitehead) [8]

Let K be the 2-complex $K = \{A \mid R\}$. Then $\pi_2(K, K^1)$ is isomorphic to the free crossed F-module with basis set R where $F = \pi_1(K^1)$.

Proof. Let C be the free crossed F-module on R constructed above. Associated with a 2-cell r is the characteristic map $\phi_r : D^2, S^1, 1 \longrightarrow K, K^1, e$. Let $\psi(r)$ be the associated class in $\pi_2(K, K^1)$. Because C is free ψ can be extended to a morphism of crossed F-modules $\psi : C \longrightarrow \pi_2(K, K^1)$. We will now show that ψ is an isomorphism.

1. *ψ is onto.*

Let an arbitrary element of $\pi_2(K, K^1)$ be represented by a disc picture p. Let $D_1, D_2, \ldots, D_k$ be the 2-cell discs in p representing the elements $r_1, r_2, \ldots, r_k$ of $R \cup R^{-1}$. Join the base point tag of D_i to the base point 1 in S^1 by a path γ_i satisfying

(i) γ_i avoids the other D_j, $j \neq i$;

(ii) γ_i meets the 1-cell arcs transversely; and

(iii) γ_{i+1} is below γ_i, $i = 1, 2, \ldots, k-1$, see Figure 18.

Then associated with γ_i in the usual way is an element f_i in F. The word f_i is read off from γ_i by its intersections with the 1-cell arcs.

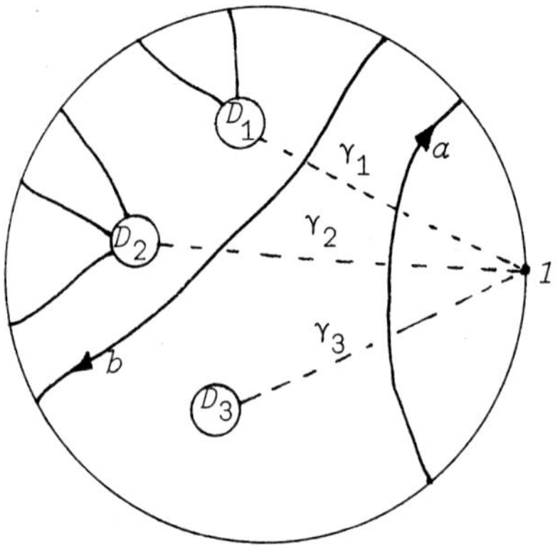

Figure 18. The associated relation string is $(f_1 \cdot r_1)(f_2 \cdot r_2)(f_3 \cdot r_3)$ where $f_1 = f_2 = ab^{-1}$ and $f_3 = a$.

Associate with this picture the relation string

$$h = (f_1 \cdot r_1)(f_2 \cdot r_2) \ \dots \ (f_k \cdot r_k).$$

Let $[h]$ denote the corresponding element in C. We will have proved that ψ is onto if we can show that $\psi[h]$ is the element of $\pi_2(K, K^1)$ represented by the picture p. If $k = 1$ then $\psi[h]$ is the conjugate of a characteristic map. If $k > 1$ the result follows by a simple induction on k.

2. ψ *is* $1-1$.

The ambiguity built into the definition of h is of three types:

1. The ordering of the 2-cell discs $D_1, D_2, \dots, D_k$.
2. The paths $\gamma_1, \gamma_2, \dots, \gamma_k$.
3. The choice of the picture p.

These ambiguities dissolve when the class $[h]$ is taken in C since:

1. An interchange of two discs corresponds to a Peiffer move on the string h .

2. Any two paths γ_i and γ_i' joining D_i to 1 give rise to the same element of F provided that they are homotopic without crossing a 2-cell disc.

The ordering of the 2-cell discs means that any two paths γ_i and γ_i' differ up to homotopy by wrapping a number of times around the disc D_i . However, this does not alter the value of h because $w(r) \cdot r = r r r^{-1} = r$ in C .

3. If two pictures represent homotopic maps then one can be obtained from another by moves M1, M2 and M3. The birth of two cancelling 2-cell discs introduces a cancelling pair $(f_i \cdot r_i)(f_i \cdot r_i^{-1})$ in the word h . The death of two such discs involves the deletion of a cancelling pair $(f_i \cdot r_i)(f_i \cdot r_i^{-1})$ in h after a sequence of Peiffer moves. (To get the cancelling 2-cells adjacent.)

The other kinds of moves do not alter h .

Conversely, given an element $h = (f_1 \cdot r_1) \ldots (f_k \cdot r_k)$ of H there is a picture to which h corresponds and which gives rise to the element $\psi(h)$ in $\pi_2(K, K^1)$.

If $\psi(h)$ is trivial then h can be reduced to the empty word by Peiffer moves and the insertion and deletion of cancelling pairs $(f_i \cdot r_i)(f_i \cdot r_i^{-1})$. Hence $[h]$ is trivial in C .

2.7 Three-dimensional complexes.

A 2-complex with one 0-cell can be thought of as a group presentation $L = \{A \mid R\}$. Now suppose that 3-cells are attached to L to form the 3-dimensional complex K . Then K is defined by the attaching maps $f : S^2 \longrightarrow L$. Furthermore, if these maps are transverse to L then they are defined by a set of identities I in the relations R .

Definition 2.7.1

An **extended group presentation** $K = \{A \mid R \mid I\}$ is a group presentation $L = \{A \mid R\}$ together with a collection of relation identities I in R. Given an extended group presentation there is associated with it a 3-complex with one 0-cell. Conversely, given a 3-complex it has the homotopy type of one associated with an extended group presentation.

2.7.2 Examples.

1. If p, q are coprime integers, then

$$K = \{ a \mid r = a^p \mid (a^q \cdot r) r^{-1} \}$$

is the lens space $L(p, q)$. If $p = 5$, $q = 3$, the spherical picture associated with the identity is shown in Figure 19.

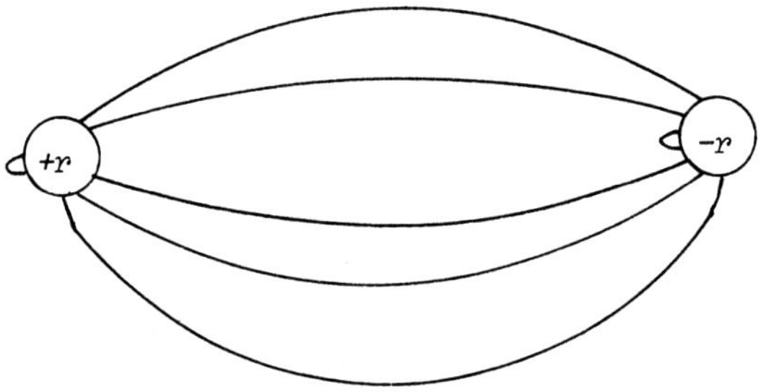

Figure 19.

2. $K = \{ a, b, c \mid r, s, t \mid r(b \cdot t^{-1}) s (c \cdot r^{-1}) \; t(a \cdot s^{-1}) \}$

where $r = [a, b]$, $s = [b, c]$, $t = [c, a]$ is the 3-torus $S^1 \times S^1 \times S^1$. For more examples of commutator identities, see the next section.

2.7.3 Identities as elements of a π-module.

If the identity multiplication is written additively the identities become elements of a module. For example, the two

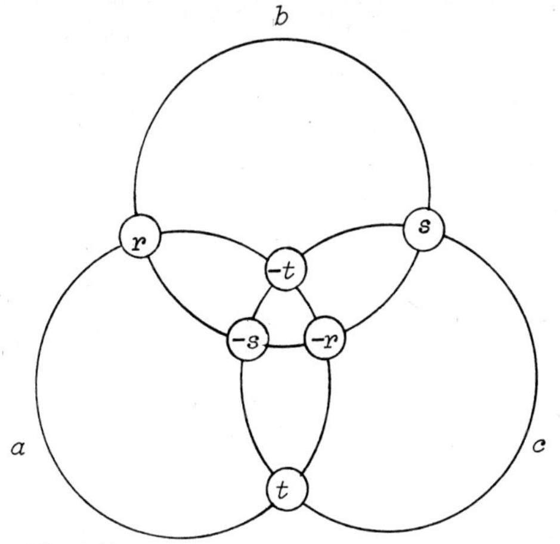

Figure 20. The identity
$[a,b](b\cdot[a,c])[b,c](c\cdot[b,a])[c,a](a\cdot[c,b]) \equiv 1$.

identities above become $(a^q - 1)\cdot r$ and $(1-c)\cdot r + (1-a)\cdot s + (1-b)\cdot t$ respectively. We will discuss this further in Chapter 4, in connection with the homology of covering spaces.

2.8 Doodles and commutator identities.

In this section the second homotopy group of the 2-skeleton of the torus $S^1 \times S^1 \times \ldots \times S^1$ is investigated using the techniques developed in the previous sections. As a consequence, certain commutator identities are exhibited.

Definition 2.8.1

A **doodle** is a collection $D = (M_1, M_2, \ldots, M_n)$ of closed compact 1-manifolds lying in the interior of the unit disc. Each manifold M, called a **colour class**, consists of a finite number of disjoint Jordan curves called **components** of D. Any three distinct components are required to be disjoint. So that

$$C_i \cap C_j \cap C_k = \emptyset \quad \text{if} \quad i \neq j \neq k \neq i.$$

If two components are coloured equally then they are disjoint and the above is automatically true. The interesting case occurs when all three are differently coloured.

Although any pair of components are not required to meet transversely, they will be drawn transverse to each other in any figurative representation.

Consider, for example, the three doodles illustrated in Figure 21.

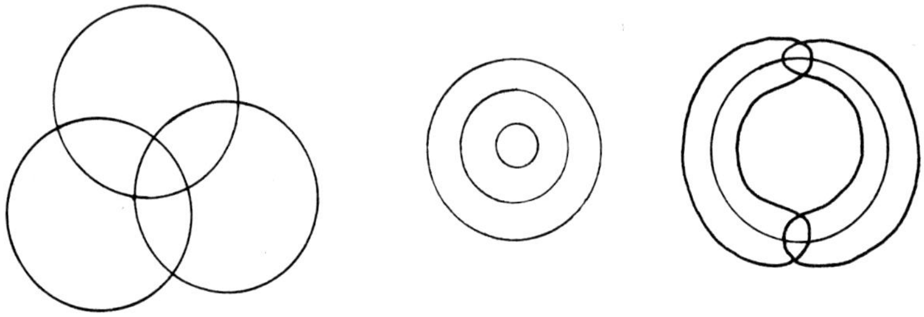

(a) (b) (c)

Figure 21. Examples of doodles.

The doodle in (a) is called the **Borromean doodle.** The doodle in (b) is a **trivial doodle.**

Doodles were introduced by Fenn and Taylor in 1977.

Definition 2.8.2 Geotopy and Isotopy.

The doodles D and D' are **geotopic** if there is a homeomorphism of D^2 which takes the components of D into the components of D'. We shall assume that the elements D are orientated and that the homeomorphism respects this orientation.

The notion of geotopy is too strict for many of our purposes. A weaker notion is that of isotopy. Two doodles D and D' are

isotopic if there is a continuous family of doodles $F(t) = (M_1(t), M_2(t), \ldots, M_n(t))$ such that $F(0) = D$ and $F(1) = D'$, $0 \le t \le 1$.

So D can be continuously varied so that it becomes D' without ever introducing a triple point. Notice that the definition also implies that components of the same colour stay disjoint during the isotopy. In Figure 21 the doodles (b) and (c) are isotopic (provided the right orientation and colourings are given), and we shall see soon that neither is isotopic to (a).

Definition 2.8.3 **Whitney moves.**

A particular kind of isotopy called a **Whitney move** is illustrated in Figure 22. During this move the number of double points is either increased or diminished by 2 .

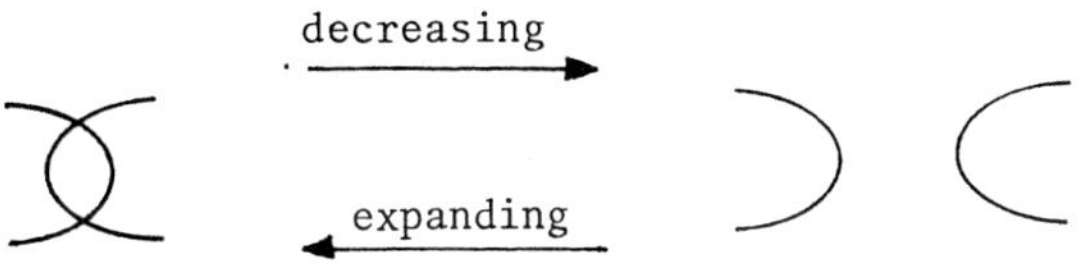

Figure 22. **Whitney moves.**

The following lemma is easy to prove.

Lemma 2.8.4 *If D and D' are isotopic doodles whose components meet transversely, then they are isotopic through a sequence of Whitney moves.*

Definition 2.8.5 **Floating components.**

A component C of doodle D is **floating** if it can be made disjoint from the other components by an isotopy.

A doodle is **trivial** if all its components are floating.

Lemma 2.8.6 *Let D and D' be isotopic doodles with transversely intersecting components and with no floating components. Then D and D' are isotopic by a sequence of Whitney moves in which all the expansive moves are made last.*

Proof. Let $D' = m_k m_{k-1} \dots m_1(D)$ where $m_1, \dots, m_k$ is a sequence of Whitney moves which transforms D into D'. Suppose that m_i increases the number of double points by 2 and m_{i+1} decreases the number of double points by 2. Then the claim is that either the order of m_i and m_{i+1} can be reversed so that m_{i+1} occurs first or m_i and m_{i+1} can be cancelled and removed from the sequence of moves.

When an expansive move creates two crossing points it also creates a 2-sided disk called the **track** of the move. Similarly, the **track** of a diminishing move is the 2-sided disk which is annihilated.

If the tracks of m_i and m_{i+1} are disjoint then it is easy to interchange them.

Suppose that m_i creates the crossing points P, Q and m_{i+1} annihilates R, S.

If $\{P, Q\} \cap \{R, S\} = \{Q\} = \{R\}$ say, then we have the situation pictured below.

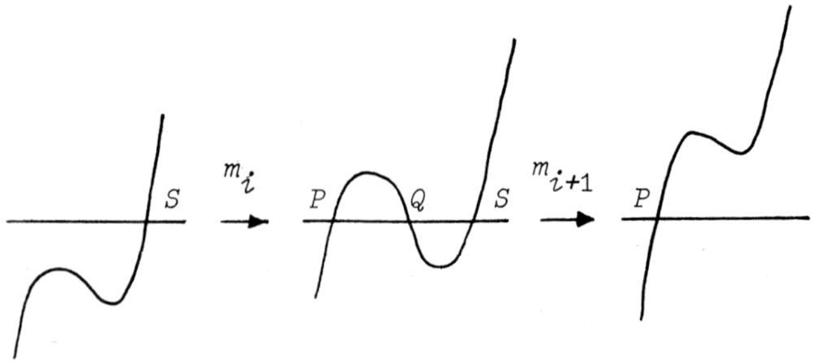

Figure 23.

In this case m_i and m_{i+1} can be cancelled. If $\{P, Q\} = \{R, S\}$ then we either have a pair of adjacent cancelling Whitney moves or the situation described in Figure 23.

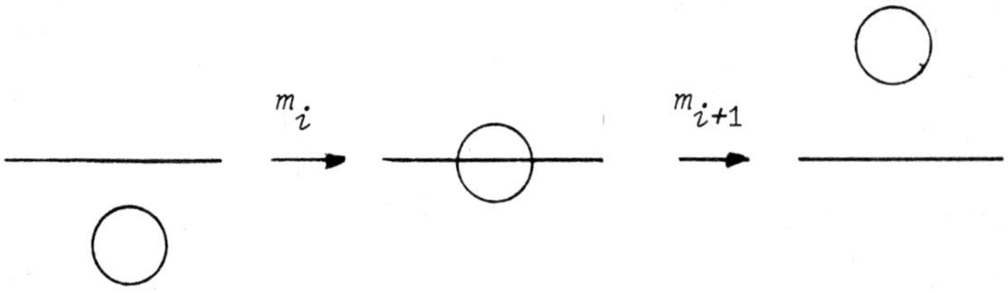

Figure 23.

However, this situation cannot occur as there are no floating components.

Definition 2.8.7 Primitive doodles.

A doodle is **primitive** if it has no floating components and is minimal with respect to the number of double points.

Corollary 2.8.8 *Isotopic primitive doodles are geotopic.*

Corollary 2.8.9 *Any doodle is isotopic to a unique primitive doodle together with a number of disjoint floating components.*

So in particular the Borromean doodle is not isotopic to the trivial doodle.

2.9 Commutator identities.

Let α be a path in D^2 which meets the components of a doodle transversely. Then if α meets the components labelled (coloured) $x_1, x_2, \ldots, x_n$ consecutively, the word $w(\alpha) = x_1^{\varepsilon_1} x_2^{\varepsilon_2} \ldots x_n^{\varepsilon_n}$ can be read off using the usual left-hand rule for the signs ε_i, [9].

2.9.1 Examples.

In the following examples the doodle components are labelled a, b, c etc. and the initial point of any reading path is chosen suitably.

1. Consider the Borromean doodle. For any component C, $w(C) = [a, b]$.

2. For the doodle in Figure 24, $w(C) = 1$, after cancellation.

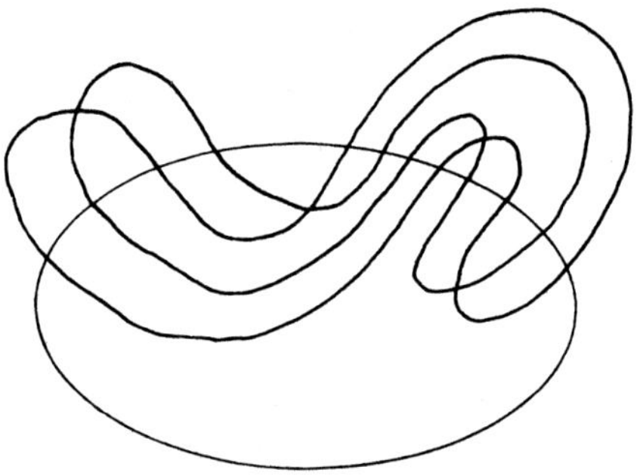

Figure 24. A doodle with $\mu = 0$.

3. For the doodle in Figure 25, $w(C) = [a, [b, c]]$.

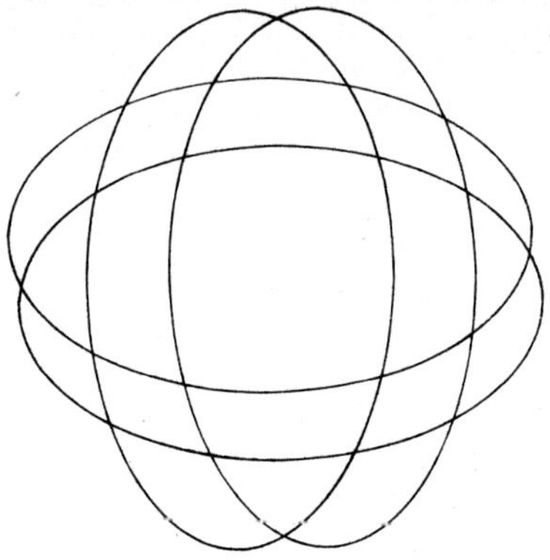

Figure 25.

2.9.2 The μ - invariant.

The words read round the components of a doodle with three components are all of the form $[a, b]^{\mu}$ for some integer μ. More specifically, let $D = (C_1, C_2, C_3)$ be a doodle. Label the components C_1, C_2 with a, b

Theorem 2.9.3 *With suitable orientations of* C_1, C_2, *there exists an integer* μ *such that* $w(C_3) = [a, b]^{\mu}$ *after cancellation and cyclic permutation.*

Proof. Consider Figure 26 below. The component C_3 is moved by an illegal move to C'_3 which has one less crossing of C_1 and C_2 in its interior [10].

Then up to cancellation and cyclic permutation of $w(C_3)$ and $w(C'_3)$, $w(C'_3)\,[a, b]^{\pm 1} = w(C_3)$.

Now proceed by induction on the number of crossing points inside C_3.

In general, if the number of colour classes is three but the number of components is unlimited, then μ can still be defined as the sum of the various μ associated with each component of the appropriate colour. So let $D = (M_1, M_2, M_3)$ be a three-

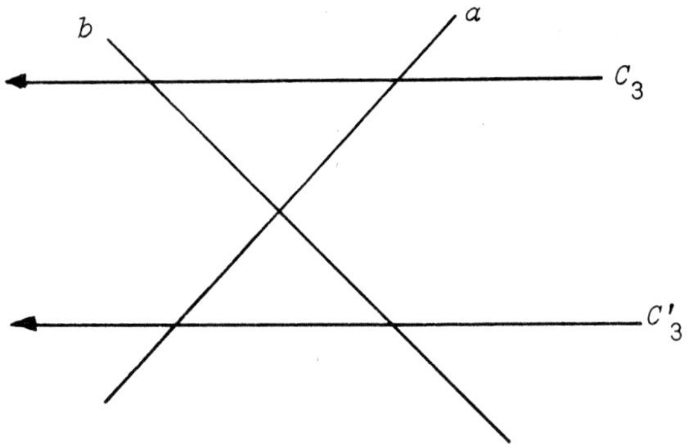

Figure 26.

coloured doodle.

Theorem 2.9.4 *With the notation above,* μ *is the algebraic number of crossing points of* M_1 *and* M_2 *inside* M_3 *where*

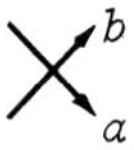

is a positive crossing point and

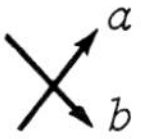

is a negative crossing point.

Proof. Similar to 2.9.3.

The integer μ is clearly an isotopy invariant and also enjoys certain symmetry properties which we shall investigate later.

Doodles exist for which μ can take any value. For example, the doodle below has $\mu = \pm 3$.

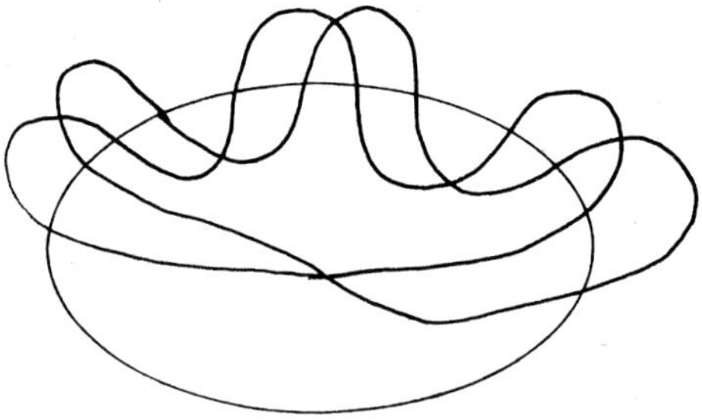

Figure 27. A doodle with $\mu = \pm 3$.

However, the invariant μ is not a complete isotopy invariant. For example, the doodle in Figure 24 is non-trivial but has $\mu = 0$.

2.9.5 Identities.

If C is a floating component in the doodle D then $w(C)$ will be the trivial word after cancellation. (Although the converse is not true.) This means it is possible to read off interesting commutator identities from a doodle. Conversely, a commutator identity gives rise to a doodle, although not uniquely. To see this, consider an identity

$$\iota = (f_1 \cdot c_1) \ldots (f_k \cdot c_k) \equiv 1$$

where the f_i lie in the group $F(a_1, \ldots, a_n)$ and the c_i are commutators. Recalling the method of relating a relation string to a disc picture, construct part of a doodle meeting a k flowered plant, as follows:

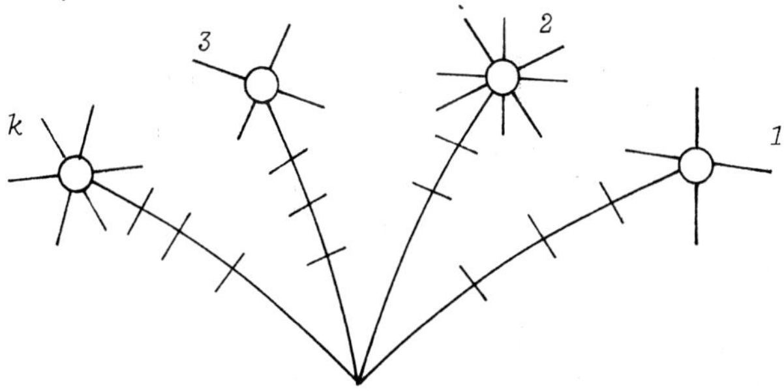

The plant stems cut the doodle arcs in the words f_i and the flower discs are spiders for the commutators c_i. Inside the discs the doodle arcs may be joined up coherently, but with a number of ambiguous crossings. Outside the plant the doodle arcs may be joined up coherently, without crossings, to form a complete doodle, since ι is an identity.

We have already seen the identity

$$[a, b]\, b\, [a, c]\, b^{-1}\, [b, c]\, c\, [b, a]\, c^{-1}\, [c, a]\, a\, [c, b]\, a^{-1} \equiv 1,$$

in Section 2.7. Here are some more examples. The floating component is shown dotted.

2.9.6 Examples of identities.

1.

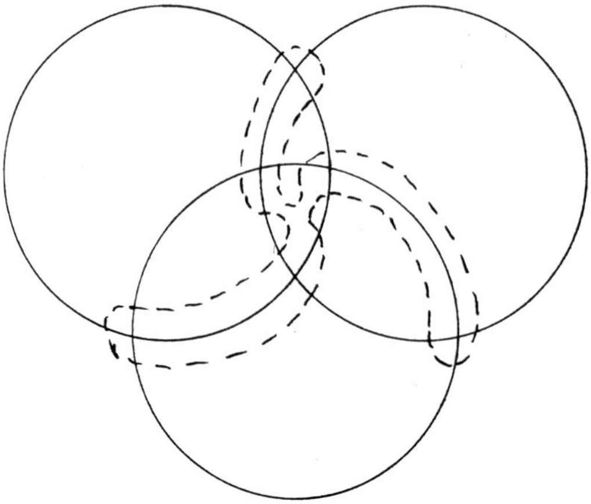

$[a, bc][b, ca][c, ab] \equiv 1$.

2.

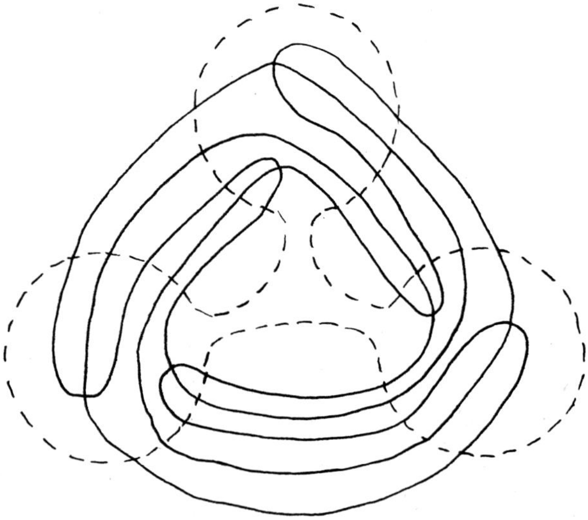

$[[a, b], bcb^{-1}][[b, c], cac^{-1}][[c, a], aba^{-1}] \equiv 1$, [11].

2.10 Concordance of Doodles.

In many ways the equivalence of isotopy is too constraining. A weaker notion is concordance, [12] .

Let $D = (C_1, \ldots, C_n)$ and $D' = (C'_1, \ldots, C'_n)$ be two doodles and assume that D lies in $D^2 \times \{0\}$ and D' lies in $D^2 \times \{1\}$.

Then D and D' are **concordant** if there are n surfaces $\left\{X_i\right\}_{i=1}^{n}$ properly embedded in $D^2 \times I$ with the following properties:

C1. Each X_i is homeomorphic to $S^1 \times I$.

C2. One boundary component of X_i is C_i and the other is C'_i.

C3. $X_i \cap X_j \cap X_k = \emptyset$ if $i \neq j \neq k \neq i$.

This last condition implies that the surfaces have no triple points of intersection.

By placing the surfaces X_i in general position with respect to the I coordinate it can be seen that D and D' are related to each other by a sequence of Whitney moves, together with two new moves:

1. Birth or death of small floating components.

2. A **bridge move** in which appropriately ordered arcs of the same colour are interchanged, as in Figure 28.

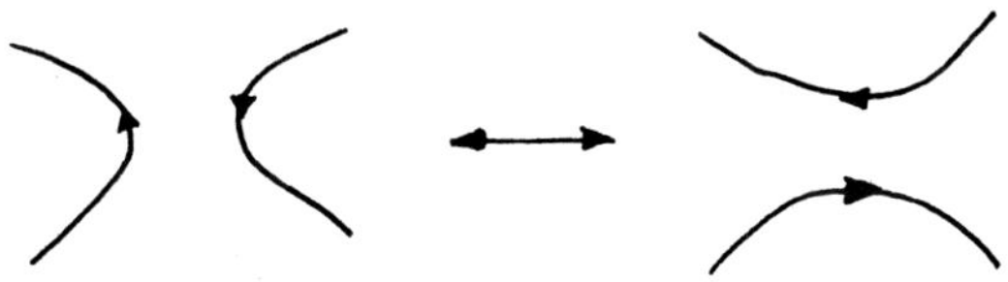

Figure 28. The bridge move.

Here is an example of a doodle which is concordant but not isotopic to the trivial doodle.

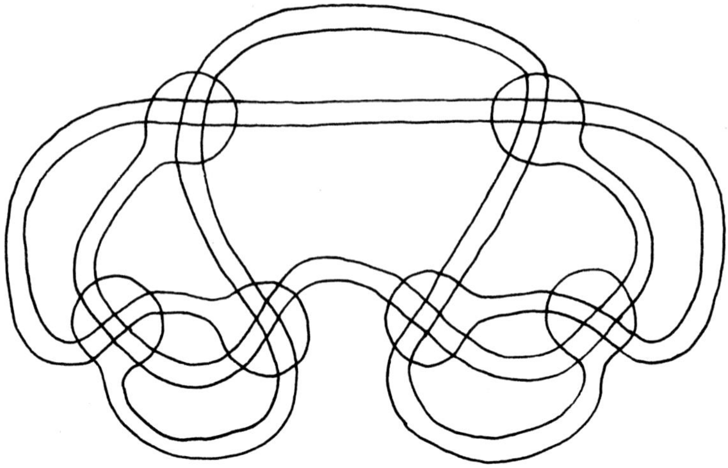

Figure 29. A non-trivial null concordant doodle.

Clearly μ is also a concordance invariant, but once again is not a complete invariant.

The doodle in Figure 24 is not null concordant but has $\mu = 0$.

2.10.1 Triple points and the μ-invariant.

Suppose a doodle $D = (C_1, C_2, C_3)$ is concordant to the trivial doodle, then if D is embedded in the boundary of the ball B^3 this means that each C_i can be spanned by discs D_i in B^3 such that $D_1 \cap D_2 \cap D_3 = \emptyset$. More generally, we can say the following:

Theorem 2.10.2 *Let $D = (C_1, C_2, C_3)$ be a doodle in the boundary of the ball B^3 and suppose that D_1, D_2, D_3 are three properly embedded discs in B^3 in general position and with $\partial D_i = C_i$, $i = 1, 2, 3$. Then the discs have at least $|\mu|$ triple points of intersection.*

Proof. The discs represent generators ξ_i of the cohomology groups $H^1(B^3, C_i) \approx \mathbb{Z}$, $i = 1, 2, 3$. The algebraic sum of the triple points represents the cup product $\xi_1 \cup \xi_2 \cup \xi_3$ in $H^3(B^3, C_1 \cup C_2 \cup C_3)$. But this product is μ times the diagonal element of $H^2(C_1 \cup C_2 \cup C_3)$.

Note that there may be triple points, even if $\mu = 0$. For example, the doodle in Figure 24 would require at least 2 triple points of opposite sign.

2.11 Cobordism of doodles.

The weakest form of equivalence of doodles is called **cobordism**. This is the equivalence freely generated by the three kinds of move considered so far: Whitney moves, the birth and death of floating components and bridge moves.

This equivalence can also be interpreted as a concordance in which the surfaces X_i homeomorphic to $S^1 \times I$ are replaced by arbitrary oriented surfaces Y_i with upper boundaries $\partial_+ Y_i = C'_i$ and lower boundaries $\partial_- Y_i = C_i$ where $\partial Y_i = \partial_+ Y_i \cup \partial_- Y_i$. The induced orientations on the boundaries will of course have to agree with the orientations on the C_i and C'_i.

The integer μ is still an invariant and this time if $\mu = 0$ the doodle is cobordant to the trivial doodle. For example:

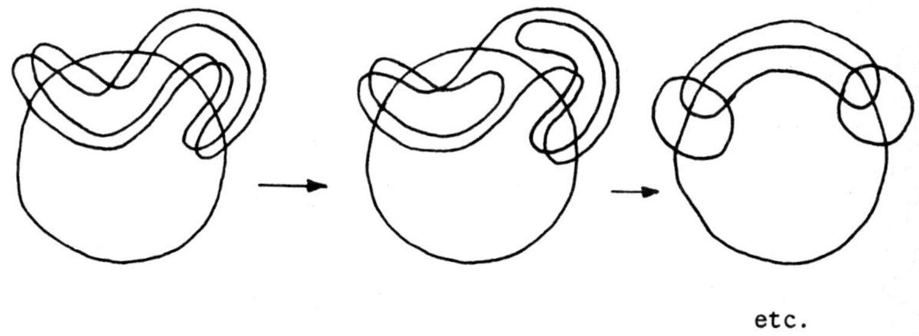

Figure 30.

In a similar manner to 2.10.2 we can prove the following:

Theorem 2.11.1 *If $D = (M_1, M_2, M_3)$ lies in the boundary of B^3 and is spanned by properly embedded oriented surfaces in general position, then μ is the algebraic number of triple points of intersection of these surfaces.*

Let Cob_n denote the set of cobordism classes of doodles containing at most n 1-manifolds, and let $[D]$ denote the cobordism class of the doodle D in Cob_n. Assume that doodles are coloured with the colours $a_1, a_2, \ldots, a_n$, and let A^n be the free abelian group, written multiplicatively and with basis $\{a_1, a_2, \ldots, a_n\}$.

Definition 2.11.2 Cob_n **as an** A^n **module.**

If D_1 and D_2 are two doodles let their **sum** be the disjoint union as in Figure 31.

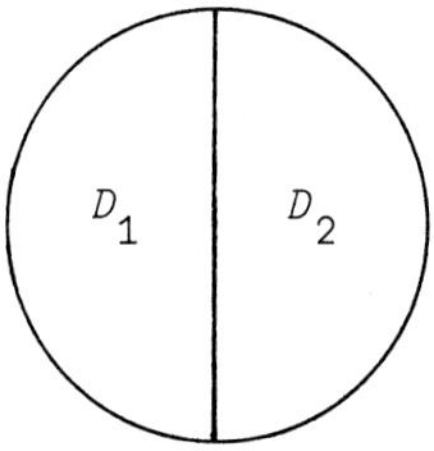

Figure 31. The sum of two doodles.

It is simple to see that this sum respects cobordism, so that $[D_1] + [D_2] = [D_1 + D_2]$ is well defined. Moreover, the fact that $[D_1] + [D_2] = [D_2] + [D_1]$ is immediate.

The empty doodle acts as a zero and the inverse $-D$ of a doodle is obtained by reflecting D in a disjoint line and then reversing the component orientations, see Figure 32.

With these definitions Cob_n becomes an abelian group. The action of a_i on D is illustrated in Figure 33.

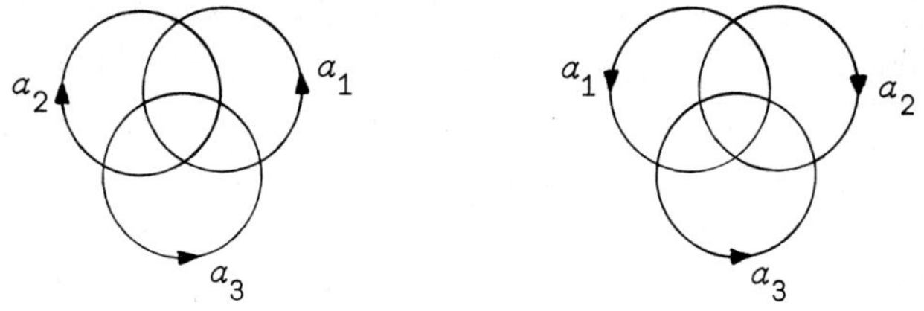

Figure 32 A doodle and its inverse.

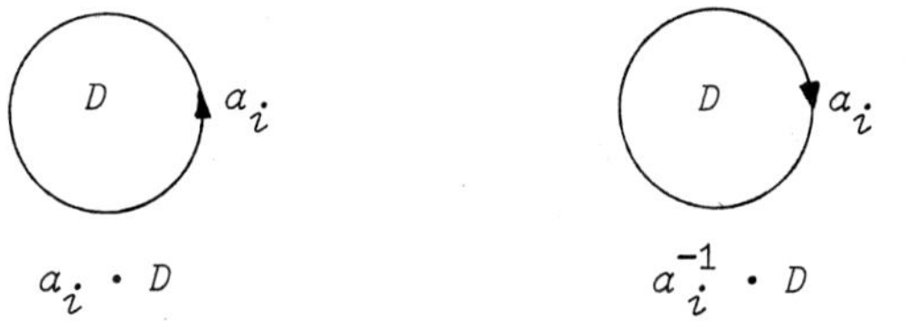

Figure 33 The action of A^n on D .

This means that the element M_i' of $a_i \cdot D\,(a_i^{-1} \cdot D)$ coloured with a_i has one more component than the element of D coloured with a_i . The new component encircles D anticlockwise (clockwise).

The action of a general element of A^n on Cob_n which makes it into an A^n *module* is defined by repetition.

Let E_n be the two-dimensional cell complex with n 1-cells and $\binom{n}{2}$ 2-cells given by

$$E_n = \left\{ a_1, \ldots, a_n \mid [a_i, a_j] = 1, \; 1 \le i < j \le n \right\}.$$

Then $\pi_1(E_n) = A^n$ and $\pi_2(E^n)$ is an A^n *module.*

Theorem 2.11.3 *Cob_n and $\pi_2(E_n)$ are isomorphic A^n modules.*

Proof. Using the theory of spherical pictures developed earlier, it is clear that any transverse map : $(S^2, 1) \longrightarrow (E_n, e)$ gives rise to a doodle, and conversely. Moreover, two maps are homotopic if and only if their corresponding doodles are cobordant.

Definition 2.11.4 **The generators of Cob_n.**

Of particular interest in Cob_n are the Borromean doodles B_{ijk} shown in Figure 34.

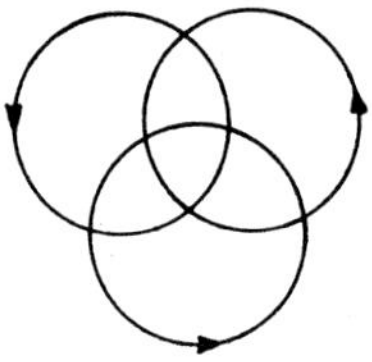

Figure 34. B_{ijk}, $i \neq j \neq k \neq i$.

Then $[B_{ijk}] = [B_{jki}] = [B_{kij}]$ and B_{ijk} and B_{ikj} are inverse to each other in Cob_n.

Notice also that B_{ijk} and B_{ikj} are not cobordant, since their μ-invariants have opposite sign.

Other possibilities can be obtained by reversing the orientation of the components coloured a_{ic}, a_j, a_k. However, these resulting doodles are all translates of the original B_{ijk} or B_{ikj}. For example,

$$[B_{-ijk}] = a_i^{-1} \cdot [B_{ijk}]$$

$$[B_{i-j-k}] = (a_j a_k)^{-1} \cdot [B_{ijk}]$$

Finally, we can extend the definition of B_{ijk} to all triples ijk by putting $B_{ijk} = 0$ if two of ijk are equal.

Theorem 2.11.5 *Every element of* Cob_n *can be written as a sum* $\sum \lambda_{ijk} [B_{ijk}]$ *where* λ_{ijk} *is an element of the group ring* $\mathbb{Z}[A^n]$.

Proof. In general, the interior of a component C of a doodle D will contain spanning arcs and other components disjoint from C. By a series of bridge moves and Whitney moves we can assume that all interiors contain either:

1. two spanning arcs which cross; or
2. other disjoint components.

The resulting doodle is of the required form. Unfortunately, if $n > 3$ this sum will not be unique. We shall give the exact result in Chapter 4.

In the meantime, the above theorem allows us to prove the promised symmetry theorem for the integers μ.

Let $D = (M_1, M_2, M_3)$ be a doodle with three elements and let $\mu(1, 2, 3)$ be the μ-invariant of D.

Theorem 2.11.6 *With the above notation, the following symmetry properties are satisfied by* μ *:*

1. $\mu(1, 2, 3) = -\mu(2, 1, 3)$;
2. $\mu(1, 2, 3) = \mu(2, 3, 1)$.

Proof. First write the cobordism class of D as an A^n-*linear* combination of Borromean doodles. Then since 1. and 2. are satisfied by the Borromean doodles, the same is true for D by linearity.

2.12 Group isomorphisms and homotopy equivalences.

In this section we look at the relationship between two

2-complexes with isomorphic fundamental group.

Definition 2.12.1 **Tietze Moves.**

Let $K = \{A \mid R\}$ be a presentation of a group π. In what follows let s be a consequence of the relations R in $F(A)$. Let x be a new generator disjoint from A and let w be an element of $F(A)$.

The **Tietze moves** T_1, T_1', T_2, T_2' act on K as follows:
$T_1\ (T_1')$. Add (delete) a generator and a relation which expresses that generator as a word in the other generators. i.e.

$$\{A \mid R\} \longleftrightarrow \{A \cup x \mid R \cup xw^{-1}\}.$$

$T_2\ (T_2')$. Add (delete) a relation which is a consequence of the other relations. i.e.

$$\{A \mid R\} \longleftrightarrow \{A \mid R \cup s\}.$$

Theorem 2.12.2 **(Tietze)** *If* $K = \{A \mid R\}$ *and* $L = \{B \mid S\}$ *are group presentations, then they present the same group if and only if they are related by a sequence of Tietze moves.*

Proof. See [13].

As an example of the above theorem, consider the presentations $\{a, b, c \mid abc = bca\}$, $\{a, b, x \mid ax = xa\}$, and the following Tietze moves between them.

$$\{a, b, c \mid abc = bca\} \xrightarrow{T_1} a, b, c, x \mid abc(bca)^{-1}, x(bc)^{-1}\}$$

$$\xrightarrow{T_2} \{a, b, c, x \mid ax(xa)^{-1}, x(bc)^{-1}, abc(bca)^{-1}\}$$

$$\xrightarrow{T_2'} \{a, b, c, x \mid ax(xa)^{-1}, x(bc)^{-1}\}$$

$$\xrightarrow{T_2} \{a, b, c, x \mid ax(xa)^{-1}, c(b^{-1}x)^{-1}, x(bc)^{-1}\}$$

$$\xrightarrow{T_2'} \{a, b, c, x \mid ax(xa)^{-1}, c(b^{-1}x)^{-1}\}$$

$$\xrightarrow{T_1'} \{a, b, x \mid ax(xa)^{-1}\}.$$

The importance of Tietze's theorem is that it shows that an isomorphism between groups can be reduced to a sequence of fairly simple moves. So that any property of presentations which is unaltered by Tietze moves is necessarily a property of the group itself.

On the other hand, given two presentations of the same group, in general it may be difficult if not impossible to actually find a sequence of Tietze moves joining them.

Here is one consequence of Tietze's theorem.

Theorem 2.12.3 *Let K, L be 2-complexes presenting the same group. Then for certain integers m and n*

$$K \vee S^2 \vee \ldots \vee S^2 \ (m \text{ terms}) \simeq L \vee S^2 \vee \ldots \vee S^2 \ (n \text{ terms}) \ [14]$$

Proof. Since K and L present the same group they are related by a sequence of Tietze moves. Now T_1 and T_1' leave the homotopy type unaltered [15].

Moreover, it is easily seen that T_2 and T_2' may be replaced by:

S_1. Replace a relator r by $r = wsw^{-1}$, where $s \in R \cup R^{-1}$ and w is an arbitrary word in $F(A)$.

$S_2\,(S_2')$. Add (delete) the relation $1 = 1$. The move S_1 is of the general sort which alters the attracting maps of the relation 2-cells by a homotopy. So that S_1 does not alter the homotopy type of the complex [16].

Finally, note that $S_2\,(S_2')$ adds (deletes) the one point union with a sphere.

Examples due to Dunwoody and Metzler show that there are complexes K, L with $K \vee S^2 \simeq L \vee S^2$ but $K \not\simeq L$. So the move S_2 or S_2' may well be necessary.

2.12.4 Nielsen Transformations.

The Nielsen transformations are very important in group theory and should be known to every topologist.

Let $K = \{A \mid R\}$ present the group π. Then the **Nielsen moves** are :

N_1. Replace some $r \in R$ by its inverse r^{-1}.

N_2. Replace r by rs where $r, s \in R$, $r \neq s$.

If to this list we add the move:

AC. Replace r by any conjugate wrw^{-1} in $F(A)$;

then we are in a position to state the following:

2.12.5 The Andrews-Curtis conjecture.

If $\{A \mid R\}$ *and* $\{B \mid S\}$ *are presentations of the trivial group with* $\#(A) = \#(B)$, $\#(R) = \#(S)$, *then the two are related by a sequence of Nielsen moves and* AC.

Notes on Chapter 2.

1. Invented by Hurewicz in 1935.

2. $\pi_1(X)$ is often denoted by $\pi(X)$ and called the fundamental group.

3. This presentation is due to Higman in 1951. Remember that $[a, b] = aba^{-1}b^{-1}$.

4. The Whitehead Conjecture is that if K is 2-dimensional and $\pi_2(K) = 0$ (aspherical), then for any subcomplex L of K, $\pi_2(L)$ is also zero.

Another unsolved conjecture is the following: given a presentation K of a non-trivial group is it possible to add a generator and relation (1-cell and 2-cell) so that the result is a presentation of the trivial group? This conjecture is sometimes called the Kervaire conjecture.

5. Remember that these homotopy classes are taken relative to K^1.

6. The group F need not be free for the definition to work, but in all cases in the book it will be.

7. Invented by Peiffer in 1949.

8. See also Brown 1980.

9. Remember that the components are oriented.

10. An inward pointing normal points to the left.

11. This identity is due to P. Hall 1933.

12. Called cobordism in Fenn and Taylor.

13. Tietze 1908 or there is a proof in Crowell and Fox.

14. Proved by Whitehead in 1939.

15. In the language of simple homotopy theory these are expansions and collapses.

16. Nor indeed the simple homotopy type.

3: COVERING SPACES

'Somehow it managed to be in two different places at once, on different levels'

Robert A. Heinlein - And he built a crooked house -

The theory of covering spaces arose from Riemann's work in defining spaces, now called **Riemann Surfaces**, on which multi-valued functions become single valued, [1].

An equivalent notion is to construct the graph of a multi-valued function.

Consider as an example the function of a complex variable, $arg\ z$. If $z = 0$, $arg\ z$ has no meaning so restrict attention to $\mathbb{C}^*$ the complex plane with the origin removed. Let $\tilde{\mathbb{C}}^*$ be the subset of R^3 consisting of all points $(z, arg\ z)$. Suppose $arg\ z$ takes its principle value θ where $0 \leq \theta < 2\pi$. Then all other values of $arg\ z$ are $\theta + 2n\pi$ where n is any integer. On travelling round the origin and back to z, $arg\ z$ changes by $\pm 2\pi$ with the positive sign for an anticlockwise motion.

It is not hard to see that $\tilde{\mathbb{C}}^*$ is an infinite spiral or helix wrapping round the *vertical axis* in a right-handed fashion, [2].

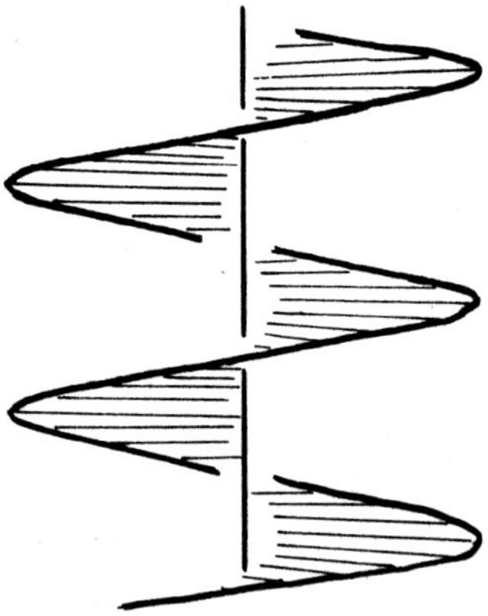

Figure 1. $\tilde{\mathbb{C}}^*$

Consider the following properties satisfied by $\tilde{\mathbb{C}}^*$.

1. There is a map $p : \tilde{\mathbb{C}}^* \longrightarrow \mathbb{C}^*$ given by $p\,(z, \arg z) = z$.

2. Given z in $\mathbb{C}^*$ there is a neighbourhood U of z such that $p^{-1}U$ consists of an infinite number of disjoint copies of U, one for each integer n. For example, if U is $\mathbb{C}^*$ with the positive x-axis removed, then $p^{-1}U$ is the disjoint union of the sets

$$\tilde{U}_n = \{(z, \theta) \mid \theta \in \arg z \quad \text{and} \quad 2n\pi < \theta < (2n+1)\pi\}.$$

3. There is an action of $\mathbb{Z}$ on $\tilde{\mathbb{C}}^*$ given by $n \,.\, (z, \theta) = (z, \theta + 2n\pi)$ with $p\,(n \,.\, z) = p\,(\tilde{z})$ for each $\tilde{z}$ in $\tilde{\mathbb{C}}^*$.

4. The space $\tilde{\mathbb{C}}^*$ is simply connected. That is, it is connected and $\pi_1(\tilde{\mathbb{C}}^*) = 1$.

As another example, consider the multivalued function $z^{\frac{1}{2}}$ also defined on $\mathbb{C}^*$, [3]. Its graph Γ lies in $\mathbb{C} \times \mathbb{C}$ and is more difficult to visualise. However, it is almost the same as $\tilde{\mathbb{C}}^*$ except that two twists around the spiral are the same as no twists.

The salient properties enjoyed by Γ are:

1. As before, there is a map $p : \Gamma \longrightarrow \mathbb{C}^*$.

2. For each z in $\mathbb{C}^*$ there is a neighbourhood U such that $p^{-1}U$ consists of the disjoint union of two copies of U, U_+ and U_-. For example, if U consists of $\mathbb{C}^*$ with the positive x axis removed, then

$$U_+ = \{(z, \zeta) \mid z \in U, \ \zeta^2 = z \quad \text{and} \quad 0 < \arg \zeta < \pi\},$$

$$U_- = \{(z, \eta) \mid z \in U, \ \eta^2 = z \quad \text{and} \quad \pi < \arg \eta < 2\pi\}.$$

3. There is an action of $\pi_1(\mathbb{C}^*)/2\pi_1(\mathbb{C}^*) = \mathbb{Z}_2$ on Γ given by $t \,.\, (z, z^{\frac{1}{2}}) = (z, -z^{\frac{1}{2}})$ where t is the generator of $\mathbb{Z}_2$.

4. The space Γ is connected, $\pi(\Gamma) = \mathbb{Z}$ and the induced map $p_* : \pi_1(\Gamma) \longrightarrow \pi_1(\mathbb{C}^*)$ is multiplication by 2.

The properties satisfied by $\mathbb{C}^*$ and Γ can be used as the basis of a definition.

Definition 3.1.1

Let X be a space. A **cover** of X consists of a space $\tilde{X}$, the **covering space** and a map $p : \tilde{X} \longrightarrow X$, the **projection**, such that X is the union of open sets U, called **fundamental sets** or **regions**, such that each component $\tilde{U}$ of $p^{-1}U$ is open in $\tilde{X}$ and $p \mid \tilde{U} : \tilde{U} \longrightarrow U$ is a homeomorphism, [4].

The space $\tilde{X}$ is sometimes referred to as the **upstairs space** and X as the **downstairs space**.

For each x in X, $p^{-1}x$ is called the **fibre** over x. It will always be a discrete space.

If X and $\tilde{X}$ are connected, then the cover is called **connected**. All covers will be connected unless specifically stated.

Covering spaces can be given this interpretation. Suppose an ant lived in X and was suddenly transported to $\tilde{X}$. Then there would be no internal way that he could discover this fact. He could go for a long walk and come back to an exact copy of his home but, unless he had an instrument to measure the value of some external function like height or logarithm, then he would have no way of distinguishing his home from all the various copies.

Examples.

1. Let F be a discrete space. Then the projection $X \times F \longrightarrow X$ is a trivial example of a disconnected cover. If U is a fundamental set of the cover $p : \tilde{X} \longrightarrow X$, then it will be shown later that $p \mid p^{-1}U : p^{-1}U \longrightarrow U$ is an example of such a cover.

2. Let $R \longrightarrow S^1$ be defined by $\theta \longrightarrow e^{i\theta}$. This is essentially the same covering space as that associated with $arg\ z$ and $log\ z$ and is an example of an infinite cyclic cover.

3. Let $S^1 \longrightarrow S^1$ be defined by $e^{i\theta} \longrightarrow e^{in\theta}$ where n is an integer. This is an example of an *n fold* cyclic cover.

4. The **real projective space** P^n is defined to be the quotient space obtained from S^n by identifying x and $-x$ for all x in S^n. The natural map $S^n \longrightarrow P^n$ is a *2-fold* cyclic cover.

5. Let $R^2 \longrightarrow S^1 \times S^1$ be defined by $(s, t) \longrightarrow (e^{is}, e^{it})$. This exhibits the plane as a cover of the torus.

6. The *3-sphere* S^3 can be thought of as pairs of complex numbers (z_1, z_2) such that $z_1\bar{z}_1 + z_2\bar{z}_2 = 1$. If p, q are coprime integers there is an action of $\mathbb{Z}_p$ on S^3 given by

$$\omega . (z_1, z_2) = (\omega z_1, \omega^q z_2)$$

where ω is a primitive p^{th} root of 1 thought of as a generator of $\mathbb{Z}_p$. The quotient space under this action is called the lens space $L(p, q)$ and the natural map $S^3 \longrightarrow L$ is a *p-fold* cyclic covering.

Notice that $L(2, 1)$ is P^3, projective space.

7. Let X be the wedge of two circles and let $\tilde{X}$ be the infinite tree pictured in Figure 2.

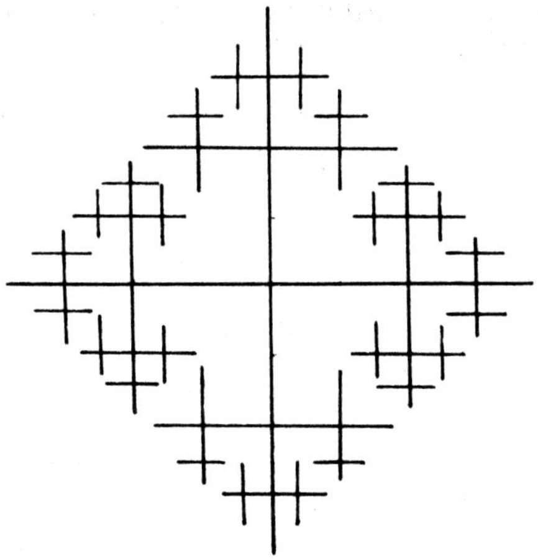

Figure 2.

A covering map $p : \tilde{X} \longrightarrow X$ is defined as follows:

Each edge of $\tilde{X}$ is either horizontal or vertical. Map the horizontal edges of $\tilde{X}$ to the first circle and the vertical edges to the second circle by exponential maps. The only points mapped to the wedge point are the vertices of $\tilde{X}$. The edges between vertices are mapped so that increasing x values (or y values) are mapped anticlockwise.

3.2 Lifting maps.

Definition 3.2.1

Let $f : Y \longrightarrow X$ be a map, then a lift of f is defined to be a map $\tilde{f} : Y \longrightarrow \tilde{X}$ such that the diagram below commutes.

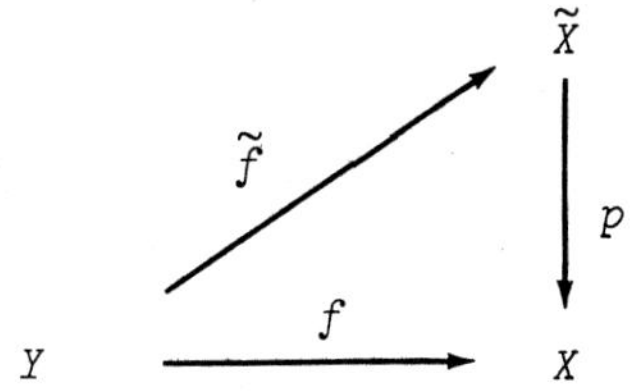

It is fortunate that for covering spaces a complete description of those maps which have lifts can be given. As for uniqueness, the following lemma is true.

Lemma 3.2.2 *Let Y be a connected space and $\tilde{f}_1, \tilde{f}_2 : Y \to \tilde{X}$ two lifts of $f : Y \to X$, then $Y_0 = \{ y \in Y \mid \tilde{f}_1(y) = \tilde{f}_2(y) \}$ is either empty or the whole of Y.*

In other words, two lifts are either the same or agree nowhere.

Proof. Let $y \in Y - Y_0$, then $\tilde{f}_1(y)$ and $\tilde{f}_2(y)$ belong to distinct components of $p^{-1}U$, where U is a fundamental set containing $f(y)$. Hence, by continuity, y has a neighbourhood in Y which lies in $Y - Y_0$. So Y_0 is closed. It only remains to show that Y_0 is also open.

Let V be a fundamental set and let $z \in Y_0 \cap f^{-1} V$.

Then $\tilde{f}_1(z) = \tilde{f}_2(z)$ lies in $\tilde{V}$ say, where $p \mid \tilde{V}$ is a homeomorphism from $\tilde{V}$ to V. Therefore every point in a neighbourhood of z also lies in $Y_0 \cap f^{-1}V$ and so Y_0 is open.

The above lemma means that the image in $\tilde{X}$ of a base point in Y uniquely determines the lift $\tilde{f}$.

The following theorem is the first of several increasingly powerful theorems about the lifts of maps.

Theorem 3.2.3 (Unique path lifting)

Let $\alpha : I \longrightarrow X$ *be a path in* X *where* $\alpha(0) = x_0$. *Pick a point* $\tilde{x}_0 \subset p^{-1}x_0$. *Then* α *lifts to a unique path* $\tilde{\alpha} : I \longrightarrow X$ *such that* $\tilde{\alpha}(0) = \tilde{x}_0$.

Proof.

Let U_0 be a fundamental region containing x_0. If $\alpha(I) \subset U_0$ then the theorem is clearly true, since there is an upstairs set $\tilde{U}_0$ containing x_0 such that $p \mid \tilde{U}_0$ is a homeomorphism onto U_0. Now, by the compactness of I there exist fundamental regions $U_0, U_1, \ldots, U_n$ in X and a subdivision $0 = x_0 < x_1 < \ldots < x_n = 1$ of I such that $\alpha([x_i, x_{i+1}])$ is contained in U_i, $i = 0, 1, \ldots, n-1$. The theorem now follows by induction on i.

Definition 3.2.4

Suppose $\alpha : I, 0 \longrightarrow X, x_0$ is a map and $\tilde{\alpha} : I, 0 \longrightarrow \tilde{X}, \tilde{x}_0$ is a lift. Let $d\alpha = \alpha(1)$. By 3.2.2 $d\alpha$ is well defined.

Corollary 3.2.5

If X *is a path connected, all the fibres* $p^{-1}x$ *have the same cardinality.*

Proof.

Let x_0, x_1 lie in X. Join x_0 and x_1 by a path α. Then the correspondence $p^{-1}x_0 \longrightarrow p^{-1}x_1$ given by $\tilde{x}_0 \longrightarrow d\alpha$ is a bijection.

In view of this result, the common cardinal of the sets $p^{-1}x$, $x \in X$ is called the **number of sheets** of the covering.

Theorem 3.2.6 (Homotopy lifting property)

Suppose $f_t : Y \longrightarrow X$ *is a homotopy of* Y *in* X, $0 \leq t \leq 1$, *where* Y *is connected. Let* $\tilde{f}_0$ *be a lift of* f_0, *then* f_t *lifts to a unique homotopy* $\tilde{f}_t$ *of* Y *in* $\tilde{X}$ *extending* $\tilde{f}_0$.

Proof.

The definition of $\tilde{f}_t$ can be seen from Figure 3.

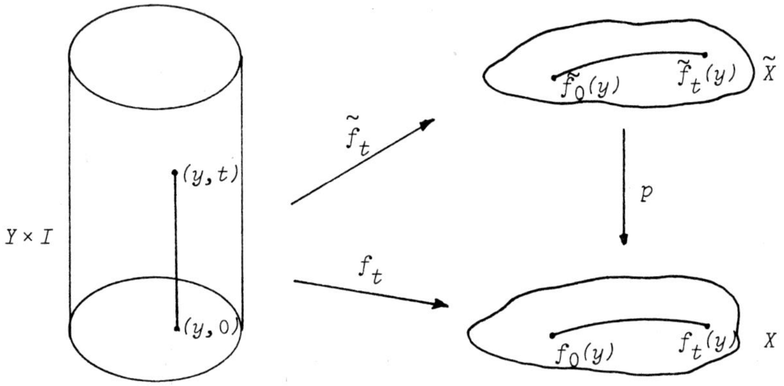

Figure 3.

In order to define $\tilde{f}_t$ just consider the path $f_s(y)$ for fixed y in Y where $0 \leq s \leq t$. This lifts uniquely with initial point $\tilde{f}_0(y)$ as was seen above. Let $\tilde{f}_t(y) = d f_s(y)$. It only remains to prove continuity.

Let (y_i, t_i) be a sequence of points in $Y \times I$ converging to (y, t). Then since $\tilde{f}_0$ is continuous, the points $\tilde{f}_0(y_i)$ converge to $\tilde{f}_0(y)$. If t is sufficiently small that the paths $\tilde{f}_s(y_i)$, $0 \leq s \leq t$ lie in the lift of a fundamental domain, then it is clear that $\tilde{f}_{t_i}(y_i)$ tends to $\tilde{f}_t(y)$. For larger t the result follows as in 3.2.3 by considering a subdivision of the path into intervals lying in such a set.

Corollary 3.2.7 *Let* α *be a loop in* X *based at* x_0. *Then the homotopy class of* α *lies in* $p_*\pi_1(\tilde{X}, \tilde{x}_0)$ *if and only if* α *lifts to a loop in* $\tilde{X}$ *based at* $\tilde{x}_0$ [5].

Corollary 3.2.8 *A covering map* $p : \tilde{X}, \tilde{x}_0 \longrightarrow X, x_0$ *induces a monomorphism* $p_* : \pi_1(\tilde{X}, \tilde{x}_0) \longrightarrow \pi_1(X, x_0)$ *and an isomorphism* $p_* : \pi_i(\tilde{X}, \tilde{x}_0) \longrightarrow \pi_i(X, x_0)$ $i > 1$.

Proof. Consider a map $\tilde{\alpha} : I^i, \partial I^i \longrightarrow \tilde{X}, \tilde{x}_0$ and a homotopy of $p\tilde{\alpha}$. This lifts to a homotopy of $\tilde{\alpha}$ so that p_* is always a monomorphism. If $i > 1$, ∂I^i is connected, so that any map $\alpha : I^i, \partial I^i \longrightarrow X, x_0$ lifts to $\tilde{\alpha} : I^i, \partial I^i \longrightarrow \tilde{X}, \tilde{x}_0$ and so p_* is onto.

Definition 3.2.9 A space Y is said to be **locally path connected** if given $y \in Y$ and a neighbourhood V of y there is a path connected neighbourhood W of y contained in V.

Theorem 3.2.10 (Lifting criterion)

Let Y *be a path connected, locally path connected space. A map* $f : Y, y_0 \longrightarrow X, x_0$ *lifts to a map* $\tilde{f} : Y, y_0 \longrightarrow \tilde{X}, \tilde{x}_0$ *if and only if* $f_*\pi_1(Y, y_0) \subset p_*\pi_1(\tilde{X}, \tilde{x}_0)$. [6].

Proof.

One way is simple. If f lifts to $\tilde{f}$ then

$$f_*\pi_1(Y, y_0) = p_*\tilde{f}_*\pi_1(Y, y_0) \subset p_*\pi_1(\tilde{X}, \tilde{x}_0).$$

Conversely, let y lie in Y. Join y_0 to y by a path α, then $f\alpha$ lifts to a unique path joining $\tilde{x}_0$ to $\tilde{f}(y)$, say.

$\tilde{f}$ **is well defined.** Suppose β is another choice for a path joining y_0 to y. Then $f(\beta\alpha^{-1})$ is a loop based at x_0 and hence lifts to a loop based at $\tilde{x}_0$, by 3.2.7.

$\tilde{f}$ **is continuous.** Let $\tilde{V}$ be an open neighbourhood of $\tilde{f}(y)$ contained in a component $\tilde{U}$ of $p^{-1}U$, where U is a fundamental region. Then $p\tilde{V} = V$ is open in U and contains $f(y)$.

Let W be a path connected open neighbourhood of y such that $f(W) \subset V$. Then $\tilde{f}(W)$ lies in $\tilde{U}$ and hence in $\tilde{V}$.

The above theorems refer to the lifting properties satisfied by a covering map. We now turn these properties on their heads and see which maps with lifting properties are covering spaces.

Definition 3.2.11

A map $p : Y \longrightarrow X$ has the **path lifting property** if, given $x \in X$, $y \in p^{-1}x$ and a path $\alpha : I \longrightarrow X$ starting at x, then there is a unique lift $\tilde{\alpha}$ of α to Y starting at y. Moreover, the lift $\tilde{\alpha}$ is required to be continuous under variations of α in the following sense:

If α_t is a continuous homotopy of α with $\alpha_t(0) = x$ and $\alpha_0 = \alpha$ then $\tilde{\alpha}_t$ is a continuous homotopy of $\tilde{\alpha} = \tilde{\alpha}_0$.

Note that this definition implies that if X is path connected then p is onto.

Theorem 3.2.12 *If $p : Y \longrightarrow X$ is open and has the path lifting property and if X is path connected and every point of X has an open simply connected neighbourhood, then p is a covering map.*

Proof.

Let U be an open simply connected neighbourhood of x in X. Let $\tilde{U}$ be a path component of $p^{-1}U$. Suppose y_0 and y_1 lie in $\tilde{U} \cap p^{-1}x$. If β is a path in $\tilde{U}$ joining y_0 to y_1 then $p\beta$ is a loop in U based at x and the unique lift of $p\beta$ is β. Now shrink $p\beta$ to x in U. By the homotopy lifting property this means that $y_0 = y_1$. So $p \mid \tilde{U}$ is a 1 - 1 continuous map from $\tilde{U}$ onto U. Since p takes open sets into open sets, $p \mid \tilde{U}$ is a homeomorphism. Therefore p is a covering map.

3.3 Classification of coverings.

The purpose of this section is to give some definition of isomorphism between coverings and classify accordingly. From now on, assume that all spaces are path connected and locally path connected.

Definition 3.3.1

Let $p_1 : \tilde{X}_1 \longrightarrow X$ and $p_2 : \tilde{X}_2 \longrightarrow X$ be two covers of X. Then a **homomorphism** from $\tilde{X}_1$ to $\tilde{X}_2$ is a map $\phi : \tilde{X}_1 \longrightarrow \tilde{X}_2$ such that the diagram

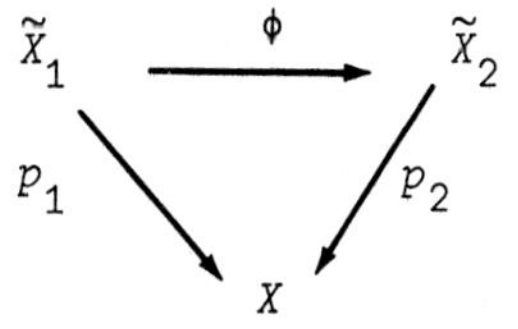

commutes.

Note that the composition of two homomorphisms is again a homomorphism and that a continuous map $\phi : \tilde{X}_1 \longrightarrow \tilde{X}_2$ is a homomorphism if and only if it takes fibres into fibres. Note also that ϕ is a lift of the covering map p_1.

Definition 3.3.2

A homomorphism $\phi : \tilde{X}_1 \longrightarrow \tilde{X}_2$ is an **isomorphism** if there exists a homomorphism $\psi : \tilde{X}_2 \longrightarrow \tilde{X}_1$ such that $\phi\psi$ and $\psi\phi$ are identity maps.

Two covering spaces are **isomorphic** if there is an isomorphism between them.

Lemma 3.3.3 *A homomorphism $\phi : \tilde{X}_1 \longrightarrow \tilde{X}_2$ is an isomorphism if and only if it is a homeomorphism.*

Lemma 3.3.4 *Let $\tilde{X}_1$ and $\tilde{X}_2$ be coverings of X and suppose $p_1(\tilde{x}_1) = p_2(\tilde{x}_2) = x_0$. Then there exists a homomorphism $\phi : \tilde{X}_1 \longrightarrow \tilde{X}_2$ such that $\phi(\tilde{x}_1) = \tilde{x}_2$ if and only if $p_{1*}\pi_1(\tilde{X}_1, \tilde{x}_1) \subset p_{2*}\pi_1(\tilde{X}_2, \tilde{x}_2)$.*

Proof. This follows from the lifting criterion 3.2.10 .

Theorem 3.3.5 *There exists an isomorphism* $\phi : \tilde{X}_1, \tilde{x}_1 \longrightarrow \tilde{X}_2, \tilde{x}_2$ *if and only if* $p_{1*}\pi_1(\tilde{X}_1, \tilde{x}_1) = p_{2*}\pi_1(\tilde{X}_2, \tilde{x}_2)$.

Proof. By 3.3.4 there are homomorphisms $\phi : \tilde{X}_1 \longrightarrow \tilde{X}_2$ and $\psi : \tilde{X}_2 \longrightarrow \tilde{X}_1$ such that $\phi(\tilde{x}_1) = \tilde{x}_2$ $\psi(\tilde{x}_2) = \tilde{x}_1$. The homomorphisms $\psi\phi$ and 1 from $\tilde{X}_1$ to itself both fix $\tilde{x}_1$ and so are equal, by 3.2.2 . Similarly, $\phi\psi = 1$.

Without specifying base points the following is true.

Theorem 3.3.6 *The covering spaces* $\tilde{X}_1$ *and* $\tilde{X}_2$ *of* X *are isomorphic if and only if the subgroups* $p_{1*}\pi_1(\tilde{X}_1, \tilde{x}_1)$ *and* $p_{2*}\pi_1(\tilde{X}_2, \tilde{x}_2)$ *are conjugate in* $\pi_1(X, x_0)$.

Proof. Suppose $\phi : \tilde{X}_1 \longrightarrow \tilde{X}_2$ is an isomorphism with inverse $\psi : \tilde{X}_2 \longrightarrow \tilde{X}_1$. Let $\phi(\tilde{x}_1) = \tilde{y}_2$, then $p_{1*}\pi_1(\tilde{X}_1, \tilde{x}_1) = $ $= p_{2*}\pi_1(\tilde{X}_2, \tilde{y}_2)$. Let $\tilde{\gamma}$ be a path in $\tilde{X}_2$ joining $\tilde{x}_2$ to $\tilde{y}_2$. Then $\tilde{\gamma}$ projects down to a loop γ in X representing some class g in $\pi_1(X, x_0)$. The group $p_{1*}\pi_1(\tilde{X}_2, \tilde{y}_2)$ is equal to the conjugate of $\pi_1(\tilde{X}_2, \tilde{x}_2)$ by g .

The converse is similar.

3.4 The Universal covering space.

In order to complete the classification of covering spaces, it is necessary to construct a covering for each subgroup of the fundamental group.

Definition 3.4.1

A covering $\tilde{X} \longrightarrow X$ is **universal** if $\pi_1(\tilde{X})$ is trivial.

In view of what has been shown above:

Theorem 3.4.2 *If a universal cover exists it is unique up to isomorphism.*

Therefore if a universal cover exists it can be called **the** universal cover.

The appellation universal follows from the following.

Theorem 3.4.3 *If* $\tilde{X} \longrightarrow X$ *is any cover and* $\tilde{X}_0 \longrightarrow X$ *is the universal cover of* X *then there is a homomorphism* $\tilde{X}_0 \longrightarrow \tilde{X}$.

Proof. Use 3.3.4 .

Not all spaces have universal covers. Consider the Hawaian earring, Figure 4. This is the union of all circles in the plane with centre $(0, -\frac{1}{2})$ and radius $1/n$.

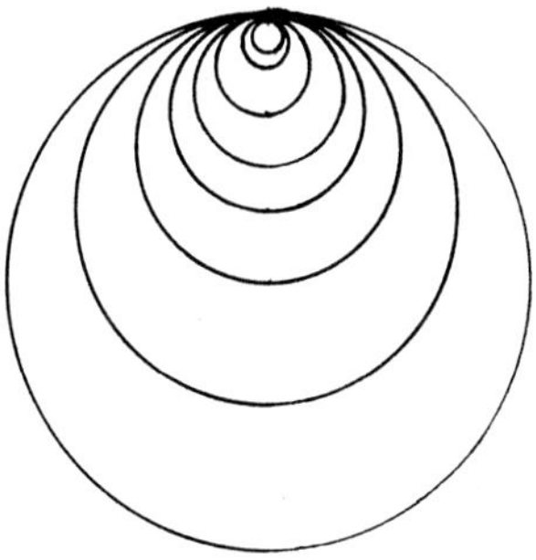

Figure 4.

Any cover of the earring must have a neighbourhood lying above the origin which is not simply connected. Accordingly, the following definition is needed.

Definition 3.4.4

A space is **locally simply connected** if every point has at least one simply connected open neighbourhood [7] .

Theorem 3.4.5 *Let* X *be a path connected locally simply connected space. Then* X *has a universal cover.*

Proof. Before giving the construction of the universal cover $\tilde{X}$ consider the following motivation for the definition if $\tilde{X}$ exists. Let $\tilde{x}_0$ be a base point above x_0. Any point $\tilde{x}$ in X can be joined by a path to $\tilde{x}_0$ and this path is unique up to homotopy keeping the end points fixed. Even more can be said; by the path lifting and homotopy lifting property, this is also true of the image of the path downstairs in X. Therefore, define $\tilde{X}$ to be the set of equivalence classes of paths in X starting at x_0. Two such paths are said to be **equivalent** if they have the same end points and are homotopic keeping the end points fixed. Let $\hat{\alpha}$ denote the equivalence class to which the path α belongs and let $p\hat{\alpha} = \alpha(1)$ be its end point, other than x_0. In this way a set $\tilde{X}$ and a map $p : \tilde{X} \longrightarrow X$ is constructed.

It remains to give a topology to $\tilde{X}$ such that p is a covering map and to show that $\tilde{X}$ is simply connected.

The topology on $\tilde{X}$

The space X is covered by open sets U which are simply connected. Call these the **fundamental regions** of X. Let α join x_0 to a point in U. Define the subset $(\hat{\alpha}, U)$ of $\tilde{X}$ to be the set of paths homotopic to α by a homotopy which keeps the final point in U.

Let U be a fixed fundamental neighbourhood of $x = p(\hat{\alpha})$, then the following statements are easily verified.

1. $\hat{\alpha} \in (\hat{\alpha}, U)$.
2. If $\hat{\alpha} \in (\hat{\beta}, U)$ then $\hat{\beta} \in (\hat{\alpha}, U)$.
3. $p\,|(\hat{\alpha}, U)$ is a bijection of $(\hat{\alpha}, U)$ onto U.
4. $p^{-1}U = \bigsqcup_i (\hat{\alpha}_i, U)$ where $\{\hat{\alpha}_i\}$ is the totality

   of all path classes in X with initial point x_0 and final point x. So $\bigsqcup\{\hat{\alpha}_i\} = p^{-1}x$.

5. The sets $(\hat{\alpha}_i, U)$ either coincide or are disjoint.

Assume a basis for the topology of X is chosen such that each element of the basis lies in a fundamental region U. Lift this basis to a basis of $\tilde{X}$ by the bijections $p\,|\,(\hat{\alpha}, U)$.

$p : \tilde{X} \longrightarrow X$ **is a cover.**

By definition and by 4. and 5., p is a continuous open map. By 5., p is a covering map.

Let α_0 denote the constant path with image x_0. If α is a path in X there is a lift $\tilde{\alpha}$ joining $\hat{\alpha}_0$ to $\hat{\alpha}$.

$\tilde{X}$ **is path connected.**

Let $\hat{\alpha}$ lie in $\tilde{X}$. For each t, $0 \le t \le 1$, there is a truncated path given by $\alpha_t(s) = \alpha(ts)$ $0 \le s \le 1$. Consider the associated path in $\tilde{X}$, $\hat{\alpha}(t) = \hat{\alpha}_t$. This joins $\hat{\alpha} = \hat{\alpha}_1$ to the constant path $\hat{\alpha}_0$. It is not hard to see that $\hat{\alpha}_t$ is continuous.

$\tilde{X}$ **is simply connected.**

Suppose α is a loop in X based at x_0 which represents a non-trivial element of $\pi_1(X, x_0)$. Then α lifts to a path $\tilde{\alpha}$ joining $\hat{\alpha}_0$ in $(\hat{\alpha}_0, U)$ to a point in the disjoint set $(\hat{\alpha}, U)$. So that only trivial loops lift to loops. Therefore $\pi_1(X, \hat{\alpha}_0) = \{1\}$, by 3.2.8.

Although this result shows the existence of universal cover, it gives no clue as to how they can be found in particular cases. This will be dealt with in the next chapter.

Definition 3.4.6

A **covering transformation** is an isomorphism $\tilde{X} \longrightarrow \tilde{X}$, between covers of X.

Note that the set of covering transformations form a group. The next result identifies this group in the case where $\tilde{X}$ is the universal cover.

Theorem 3.4.7 *The group of covering transformations of the universal cover $\tilde{X}$ can be identified with the fundamental group of X as a left group action.*

Proof. Let $\tilde{x}_0$ and $x_0 = p(\tilde{x}_0)$ be base points. Then any covering transformation is determined by the image of $\tilde{x}_0$ under the transformation.

Let $g \in \pi_1(X, x_0)$ and let α be a loop representing g. Then α lifts to a path $\tilde{\alpha}$ with initial point $\tilde{x}_0$. Let $g \cdot \tilde{x}_0$ denote its final point.

Let $\tilde{x} \in \tilde{X}$ be any other point. Join $\tilde{x}_0$ to $\tilde{x}$ by a path $\tilde{\beta}$ and let $g \cdot \tilde{\beta}$ denote the lift of $\beta = p\tilde{\beta}$ with initial point $g \cdot \tilde{x}_0$. Put $g \cdot \tilde{x}$ equal to the final point of $g \cdot \tilde{\beta}$, see Figure 5.

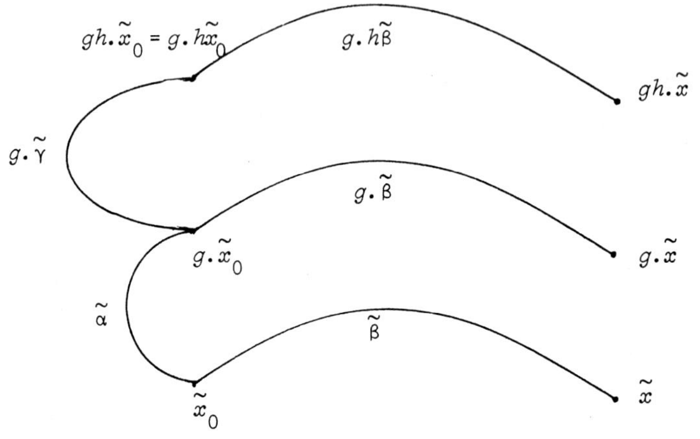

Figure 5.

Since any two paths joining $\tilde{x}_0$ to $\tilde{x}$ in $\tilde{X}$ are homotopic relative to end points, $g \cdot \tilde{x}$ is well defined.

Finally, since $g \cdot \tilde{x} = \tilde{x}$ if and only if $g = 1$, it is only necessary to check that

1. $1 \cdot \tilde{x} = \tilde{x}$ and 2. $g \cdot (h \cdot \tilde{x}) = (gh) \cdot \tilde{x}$.

Statement 1. is obvious. Statement 2. is proved as follows:

Suppose h is represented by the loop γ. Then gh is represented by the composite loop $\alpha\gamma$ and this lifts to the composition of $\tilde{\alpha}$ from $\tilde{x}_0$ to $g \cdot \tilde{x}_0$ and $g \cdot \tilde{\gamma}$ from $g \cdot \tilde{x}_0$ to $g \cdot (h \cdot \tilde{x}_0)$. So that $(g \cdot h) \cdot \tilde{x}_0 = g \cdot (h \cdot \tilde{x}_0)$. By lifting $\tilde{\beta}$ a similar result applies to $\tilde{x}$.

Examples.

1. Let X be the torus $S^1 \times S^1$, then the universal cover is the plane R^2. The fundamental group of X is the free abelian group on two generators, a and b say. The action of this group on R^2 may be represented by $a(x, y) = (x+1, y)$, $b(x, y) = (x, y+1)$. Notice that $ab(x, y) = (x+1, y+1) = ba(x, y)$, as required.

2. Consider the projective plane P^2. The universal cover is the sphere S^2 and the group of covering transformations is generated by the antipodal map $(x, y, z) \longrightarrow (-x, -y, -z)$.

3. The universal cover of the Klein bottle is the plane. The group of covering transformations is generated by two transformations at right angles. One is unit translation along the *y-axis* and the other is unit translation along the *x-axis*, followed by a flip, [8].

Now consider again the action of the fundamental group on the universal cover $\tilde{X}$ of a space X. This action is without fixed points. But much more can be said.

Definition 3.4.8

A group action G is said to be **totally disconnected** if every point has a neighbourhood U which is disjoint from all the translates gU where $g \in G$ and $g \neq 1$.

The action of $\pi_1(X)$ on $\tilde{X}$ is totally disconnected. Conversely, if G acts on Y in a totally disconnected way, then it is not hard to see that the map $Y \longrightarrow Y/G$ from Y to the space of orbits is a covering.

Definition 3.4.9

Let G denote the fundamental group $\pi_1(X, x_0)$ and let H be a subgroup. Call two points $\tilde{x}$ and $\tilde{y}$ in the universal cover

$\tilde{X}$ equivalent if $g\tilde{x} = \tilde{y}$, where $g \in H$. This is an equivalence relation, since H is a subgroup.

Let $\tilde{X}_H$ denote the space of equivalence classes with the quotient topology. There is a natural map $p_H : \tilde{X}_H \longrightarrow X$ which is a covering map and makes the diagram

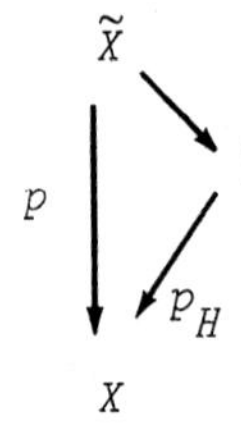

commute.

Consider a loop α in X based at x_0. This will lift to a loop $\tilde{X}_H$ if and only if the homotopy class of α lies in H, so H is the image of the fundamental group of $\tilde{X}$ under p_{H*} by 3.2.8 .

In this way a covering space unique up to isomorphism can be constructed for each subgroup of $\pi_1(X, x_0)$.

3.5 Regular covering spaces.

Definition 3.5.1

A covering $p : \tilde{X} \longrightarrow X$ is **regular** if $p_* \pi_1(\tilde{X}, \tilde{x}_0)$ is a normal subgroup of $\pi_1(X, x_0)$.

Notice that this definition does not depend on the base points chosen.

Let N be a normal subgroup of $G = \pi_1(X, x_0)$ and let $p_N : \tilde{X}_N \longrightarrow X$ be the corresponding cover and $p : \tilde{X} \longrightarrow X$ the universal cover. Then the group G/N acts on $\tilde{X}_N$ as follows. An element of $\tilde{X}_N$ can be thought of as a collection of points

$\{n\tilde{x}\}_{n \in N}$ where $\tilde{x}$ is an element of $\tilde{X}$ the universal cover

and n varies over N.

If $[g]$ is the class in G/N determined by $g \in G$ let

$$[g] \cdot \{nx\} = \{g \cdot nx\} .$$

Because N is normal, this is the same point as $\{ n_1 g x \}$, where $n_1 \in N$ and so is an element of X_N.

Any regular covering is specified by a short exact sequence of groups $1 \longrightarrow N \longrightarrow G \longrightarrow G/N \longrightarrow 1$.

The fundamental group of the cover is N and G/N is the group of covering transformations.

Important examples are $N = \{1\}$ giving the universal cover and $N = [G, G]$, the commutator subgroup, giving a cover which is often called the **universal Abelian cover.**

If M is a non-orientable manifold, then the set of orientation preserving loops gives rise to a subgroup of index 2 in the fundamental group. The corresponding double cover $\tilde{M}$ is called the **orientation cover** of M. Points of the orientable manifold $\tilde{M}$ may be thought of as a point of M together with a local orientation.

Examples.

1. Let $0 \longrightarrow \mathbb{Z} \xrightarrow{\cdot n} \mathbb{Z} \longrightarrow \mathbb{Z}_n$ be the usual representation of the integers modulo n. If $X = S^1$ the corresponding cover is the *n-fold* cyclic cover $S^1 \longrightarrow S^1$ given by $e^{it} \longrightarrow e^{int}$.

2. Imagine the Möbius band M sitting in R^3. The orientation cover is obtained by slicing M with a razor along its length to obtain a twice-twisted annulus covering M.

Definition 3.5.2

Consider the collection of all regular coverings of X. This has an **order** defined on it as follows: say $\tilde{X}_1 \leq \tilde{X}_2$ if there is a homomorphism $\phi : \tilde{X}_1 \longrightarrow \tilde{X}_2$. Under this ordering the universal covering is the initial object and the identity cover $X \longrightarrow X$ is the final object. If $\tilde{X}_1 \leq \tilde{X}_2$ and $\tilde{X}_2 \leq \tilde{X}_1$ then there are homomorphisms $\phi : \tilde{X}_1 \longrightarrow \tilde{X}_2$ and $\psi : \tilde{X}_2 \longrightarrow \tilde{X}_1$. Let the composition $\psi\phi : \tilde{X}_1 \longrightarrow \tilde{X}_1$ correspond to the element g in $\pi_1(X)$. Then if $\psi' = g^{-1}\psi$, $\psi'\phi$ is the identity. Similarly, $\phi\psi'$ is the identity. So the ordering respects isomorphism classes of regular coverings.

3.6 Irregular coverings.

Two classification theorems for covering spaces may be stated as follows:

Theorem 3.6.1 *Let X be a path connected space* [9]. *The isomorphism classes of coverings of X preserving base points are in 1 - 1 correspondence with the subgroups of* $\pi_1(X, x_0)$.

If base points are not preserved then:

Theorem 3.6.2 *Let X be a path connected space. The isomorphism classes of coverings of X are in 1 - 1 correspondence with the conjugacy classes of subgroups of* $\pi_1(X, x_0)$.

The regular coverings corresponding to normal subgroups are uniquely defined and are well behaved under the quotient group action. The irregular coverings are not nearly so well behaved and are often a source of difficulty for the beginner.

An Example.

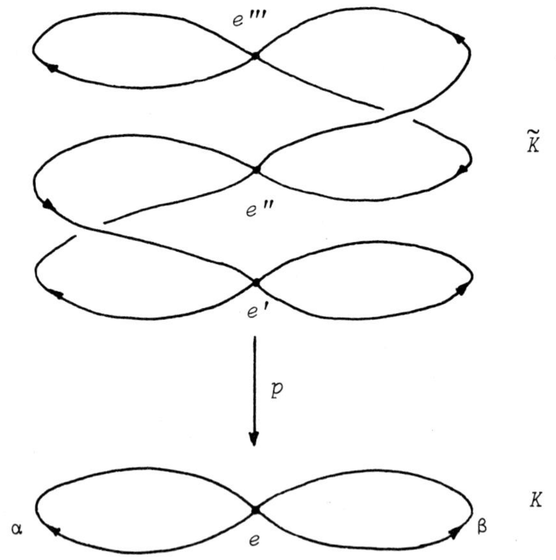

Figure 6.

Consider the covering of $K = S^1 \vee S^1$ indicated in Figure 6. The group $\pi_1(K, e)$ is $\mathbb{Z} * \mathbb{Z}$ generated by α and β. The subgroup $p_* \pi_1(K, e'')$ is the free subgroup of index 3 generated by α^2, β^2, $\alpha\beta\alpha^{-1}$ and $\beta\alpha\beta^{-1}$.

This example has two disturbing features:

1. A loop (say β) may lift to a loop at one base point (e') but may lift to a path at another base point $(e''$ or $e''')$.
2. Although $\tilde{K} \longrightarrow K$ is non-trivial, the group of covering transformations is trivial.

However, there is a map $\pi_1(K, e) \longrightarrow S_3$, the group of permutations of three objects, obtained by lifting loops and is given by:

$$\alpha \longrightarrow \begin{pmatrix} e' & e'' & e''' \\ e'' & e' & e''' \end{pmatrix}$$

$$\beta \longrightarrow \begin{pmatrix} e' & e'' & e''' \\ e' & e''' & e'' \end{pmatrix}$$

We may formulate the general case as follows:

Let $p : \tilde{X} \longrightarrow X$ be a covering. Choose base points x_0 in X and $\tilde{x}_0$ in $E = p^{-1} x_0$. Let $G = \pi_1(X, x_0)$ and let H be the subgroup $H = \pi_1(\tilde{X}, \tilde{x}_0)$ of index $n \leq \infty$ in G. As we saw in 3.4.9, the elements of E may be thought of as the equivalence classes $\{hy\}_{h \in H}$ where y lies in the universal cover over x_0. A more compact notation for an element of E is Hy. If Hy' is another element of E then $y' = gy$ for some $g \in G$ so the elements of E are all of the form Hgy and are therefore in 1 - 1 correspondence with the right cosets Hg of H in G.

Definition 3.6.3 Monodromy.

With the notation above, let $E = G/H = \{H, Hg_1, Hg_2, \ldots, Hg_{n-1}\}$ be the set of right cosets of H in G. Then there is a homomorphism m of G to the group of permutations of E given by

$$m(g) = \begin{pmatrix} H, & Hg_1, & Hg_2, & \ldots, & Hg_{n-1} \\ Hg, & Hg_1g, & Hg_2g, & \ldots, & Hg_{n-1}g \end{pmatrix}.$$

The group of permutations so obtained is called the **monodromy group** of the cover, [10]. It is written $M(\tilde{X}, \tilde{x}_0)$.

The monodromy group acts on E on the right, [11]. An alternative formulation for $m(g)$ is given by the following: Let α be a loop in X based at x_0 and representing g. Lift α to a path $\tilde{\alpha}$ with $\tilde{\alpha}(0) = \tilde{x}$, where $\tilde{x} \in E$. Then $\tilde{x} \cdot m(g) = \tilde{\alpha}(1)$.

The subgroup $H = \pi_1(\tilde{X}, \tilde{x}_0)$ is recovered by the formula $H = \{g \in G \mid x_0 \cdot m(g) = x_0\}$. In other words, H consists of those classes in $\pi_1(X, x_0)$ whose representative loops lift to loops in $\tilde{X}$ based at $\tilde{x}_0$.

Conversely, let S_n be the group of permutations of the integers $\{1, 2, \ldots, n\}$ $n \leq \infty$ and let $m : G \longrightarrow S_n$ be a homomorphism. The map m is said to be **transitive** if, for each k, $1 \leq k \leq n$ there is a $g \in G$ such that $1 \cdot m(g) = k$.

Theorem 3.6.4 *Using the notation above, there is a connected n-sheeted covering $\tilde{X}$ of X with the monodromy m if and only if m is transitive.*

Proof. An n-*sheeted* connected covering clearly has a transitive monodromy. Conversely, let $H = \{g \in G \mid 1 \cdot m(g) = 1\}$, then H has index n in G and the right cosets are $Hg_1, Hg_2, \ldots Hg_n$ where g_k is some representative of the equation $1 \cdot m(g) = k$. Let $\tilde{X}$ be the corresponding cover. Then $\tilde{X}$ is connected, as can be seen by lifting loops representing the elements g_k, $k = 1, 2, \ldots, n$.

Definition 3.6.5 A formula for the covering transformation group.

Write $A(\tilde{X})$ for the group of covering transformations of $\tilde{X}$. Then elements of $A(\tilde{X})$ act on the left on the whole of $\tilde{X}$.

(Remember that the monodromy acts on the right and only on the fibre E).

The first thing to note is that elements of $A(\tilde{X})$ are induced by the group action on the universal cover $\tilde{X}_0$. This can be seen by lifting the composition $\tilde{X}_0 \xrightarrow{p_0} \tilde{X} \xrightarrow{f} \tilde{X}$, where $f \in A(\tilde{X})$. So associated with any covering transformation is an element of the group $G = \pi_1(X, x_0)$. This group element is not unique: it depends on the choice of the base point in the fibre over $\tilde{x}_0$. Before making this dependance precise, we need the following definition from group theory.

Definition 3.6.6 The normaliser subgroup.

If H is a subgroup of G, the **normaliser** $N(H)$ of H in G is the subgroup $N(H) = \{g \in G \mid gHg^{-1} = H\}$. So $N(H)$ is the largest subgroup of G in which H is a normal subgroup.

Theorem 3.6.7 *If $\tilde{X} \longrightarrow X$ is a covering corresponding to a subgroup H of $\pi_1(X, x_0)$, then the covering transformation group $A(\tilde{X})$ is isomorphic to the quotient group $N(H)/H$.*

Proof. Pick a base point y_0 in the universal cover $\tilde{X}_0$ so that y_0 lies over $\tilde{x}_0$. If f is in $A(\tilde{X})$ then $f(\tilde{x}_0) = f(p_0 y_0) = p_0(g y_0)$, where $p_0 \colon X_0 \longrightarrow X$ is the projection and $g \in \pi(X, x_0)$. If g' is another choice of group element then $g' g^{-1}(y_0)$ lies in the fibre over $\tilde{x}_0$. So $g' g^{-1}$ is in H. In other words, the choice of g is well defined modulo H.

We shall now see that g lies in the normaliser $N(H)$. Let h be an arbitrary element of H, then y_0 and hy_0 lie in the fibre over $\tilde{x}_0$ and gy_0 and $g(hy_0)$ lie in the fibre over $f\tilde{x}_0$. So $ghy_0 = h'gy_0$ for some $h' \in H$. i.e. $gH = Hg$ and $g \in N(H)$.

The map $A(\tilde{X}) \longrightarrow N(H)/H$ given by $f \rightarrow g$ is onto because if $g \in N(H)$ there is a transformation f_g in $A(\tilde{X})$ given by

$f_g(Hy_0) = Hgy_0$ etc. Finally, the map is $1-1$ because only trivial transformations are induced by elements of H and conversely.

We now continue with three conditions, such that if any one is satisfied then the covering is regular.

Theorem 3.6.8 *The covering $\tilde{X} \longrightarrow X$ is regular if and only if one of the following is satisfied:*

1. *The group of covering transformations $A(\tilde{X})$ acts transitively.*
2. *The monodromy group $M(\tilde{X}, \tilde{x}_0)$ can be identified with the covering transformation group $A(\tilde{X})$.*
3. *If a loop based at x_0 lifts to a loop based at $\tilde{x}_0$ then it lifts to a loop at any other point of the fibre.*

Proof. This may safely be left to the reader, as may the proof of the following:

Lemma 3.6.9 *Let $p : \tilde{X} \longrightarrow X$ be a covering and suppose $\tilde{x}_0$ and $\tilde{x}_1$ lie in $p^{-1} x_0$. Let $\tilde{\alpha}$ be a path from $\tilde{x}_0$ to $\tilde{x}_1$ and let g be the homotopy class of the loop $p\tilde{\alpha}$ in X. Then $p_* \pi_1(\tilde{X}, \tilde{x}_0)$ is the conjugate subgroup $g p_* \pi_1(\tilde{X}, \tilde{x}_1) g^{-1}$ in $\pi_1(X, x_0)$.*

As an application of the above lemma, consider the complex $K = \{a, b \mid a^2 = bab^{-1}\}$ and let G be its fundamental group. Let H be the infinite subgroup of G generated by a. Then H has the curious property that bHb^{-1} is a proper subgroup of H. Let $p_H : \tilde{K}_H \longrightarrow K$ be the covering corresponding to H. Choose a base point $\tilde{e} \in p^{-1} e$ in $\tilde{K}_H$ and let $\tilde{e}_1 = db$ be the other end point of a lift of b. Then $p_* \pi_1(\tilde{K}_H, \tilde{e}) \subset p_* \pi_1(\tilde{K}_H, \tilde{e}_1)$ and so there exists a homomorphism $\phi : \tilde{K}_H \longrightarrow \tilde{K}_H$ with $\phi(\tilde{e}) = \tilde{e}_1$. However, this homomorphism is not an isomorphism.

This example shows that irregular coverings can only be ordered like the regular coverings if the coverings are considered with specified base points.

3.7 Covering spaces of surfaces.

So far, little has been said about the general theory of coverings of surfaces. To fix notation let T_γ be the closed orientable surface of genus γ and P_γ the closed non-orientable surface of genus γ.

The universal cover of the projective plane P_1 is also the orientation cover which is T_0 or the sphere S^2. The action of the fundamental group $\mathbb{Z}_2$ is given by the antipodal map.

In general, the orientation cover of P_γ is homeomorphic to $T_{\gamma-1}$ [12]. The action of $\mathbb{Z}_2$ is an orientation reversing homeomorphism without fixed points.

For $\gamma > 0$, the universal cover of T_γ and hence of $P_{\gamma+1}$ is a plane. For $\gamma = 1$ this is the Euclidean plane. There is a convex cell, a square, whose transforms under the group of covering transformations cover the plane without overlapping. Moreover, the covering transformations themselves all belong to the group of Euclidean motions.

Much the same is true for $\gamma > 1$ except that the Euclidean plane is replaced by the hyperbolic plane. This brings us to the next section.

3.7.1 Hyperbolic geometry.

This section is not meant to be a comprehensive introduction to hyperbolic geometry, but sufficient results will be introduced in order to facilitate the study of the universal cover of T_γ $\gamma > 1$.

Little or no proof will be given of statements. The interested reader will find the details in several excellent books. The following is mostly taken from Siegel and Caretheodory.

Definition 3.7.2 Setting up the model.

The model used here is due to Poincaré. Let H be the interior of the unit disc in the complex plane $\mathbb{C}$. The points of H are the *points* of the hyperbolic plane. The points on the unit circle S^1 are often called the *points at infinity*. The *hyperbolic lines*, or *H-lines* for short, are the arcs in H of circles orthogonal to S^1. Also included as H lines are diameters of H.

Two points A, B of H determine a unique H-line. Write AB for the H-arc joining them. This notation will also be used if one or both of A, B lie at infinity. If two H-lines meet at A inside H their circle extensions also meet at the inverse point $1/\bar{A}$ outside H.

If two H-lines meet only at infinity they are called parallel. If two H-lines fail to meet they are sometimes called *ultraparallel*.

If A, B, C, D are four points of C their *cross ratio* is defined to be $\delta(A, B, C, D) = (A-C)(B-D)/(A-D)(B-C)$.

Let A, B lie in H. The line through A and B meets the circle at infinity in two more points C, D : see Figure 7.

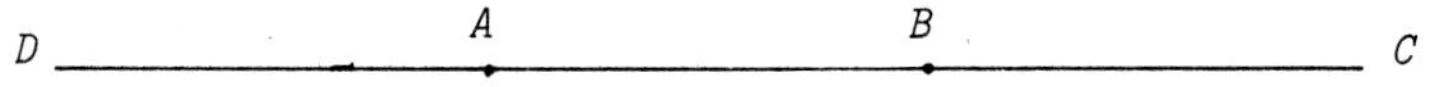

Figure 7.

The cross ratio of these points, lying on a circle, is real and positive. Define the *hyperbolic distance* between A and B by

$$d(A, B) = \log \delta(A, B, C, D).$$

In general, arc length s has an infinitesimal expression given by

$$ds = 2\,|dZ|\,/\,(1 - |Z|^2).$$

The corresponding expression for the area infinitesimal is

$$d\omega = (dZ^2 - d\bar{Z}^2)/i(1 - |Z|^2)^2 .$$

The *angle* between two H - lines is measured in this model in exactly the same way as the ordinary euclidean angle. It is an important fact in the history of non-euclidean geometry that the three angles of a triangle have sum less than π. This is a consequence of the Gauss-Bonnet theorem and the fact that the hyperbolic plane has constant negative curvature. It is also useful to consider *ideal triangles* in which one or more vertices are at infinity. In this manner polygons with angles equal to zero can be considered.

Let NM be an H - line where NM are at infinity and let A be a point of H not on the line NM. Let B be the foot of the perpendicular from A to NM. If AM is one of the H - lines through A parallel to NM and x is the distance AB let $\Pi(x)$ denote the angle BAM : see Figure 8. The function $\Pi(x)$ is called the *angle of parallelism*. For euclidean geometry this would always be $\pi/2$. Analytically $\Pi(x) = 2\tan^{-1}(e^{-x})$.

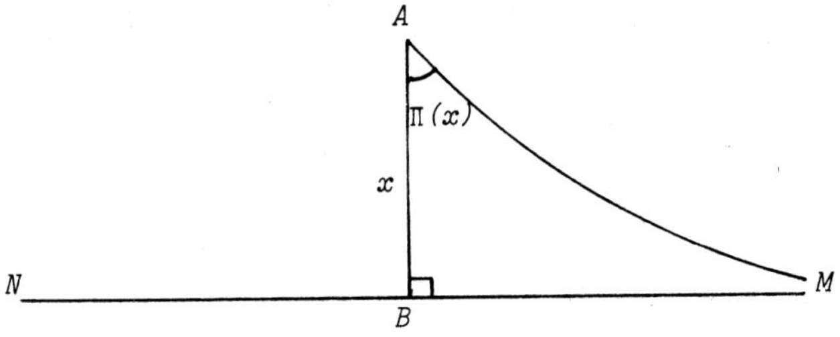

Figure 8. The angle $\Pi(x)$.

Transformations of the hyperbolic plane which preserve orientation and arc length are called (orientation preserving) isometries. They will also preserve area, angles and the sense of angles. All such isometries have the analytical form

$$Z \to W = e^{i\phi}(A - Z)/(1 - \bar{A}Z), \quad |A| < 1,$$

and hence can be thought of as acting on the whole complex plane C. They constitute the group of bilinear transformations $W = (\alpha Z + \beta)/(\gamma Z + \delta)$, $\alpha\delta - \beta\gamma \neq 0$ which preserve the unit disc, $|Z| \leq 1$ and hence its interior H. There is a unique bilinear transformation taking the points A, B, C to P, Q, R given by $(W - Q)(P - R)/(W - R)(P - Q) = (Z - B)(A - C)/(Z - C)(A - B)$. The analytic form expresses the fact that the cross ratio $\delta(Z, A, B, C)$ is preserved. Consequently there is a unique isometry of H taking two given points to two other given points the same distance apart.

The *discriminant* Δ of an isometry is given by

$$\Delta = \sin^2(\phi/2)/(1 - A\bar{A}).$$

The value of Δ divides the set of isometries into three classes.

Definition 3.7.3

Elliptic isometries, or rotations, $\Delta < 1$.

These have two fixed points ρ and $1/\bar{\rho}$ separated by the unit circle. An elliptic isometry can be thought of as a rotation about the fixed point lying inside H through an angle equal to $2\cos^{-1}\sqrt{\Delta}$. The set of orbits is the pencil of circles (and one line) with degenerate circles ρ and $1/\bar{\rho}$. The one line is the perpendicular bisector of the two fixed points : see Figure 9.

The orthogonal pencil to the orbits is permuted by the isometry.

The elliptic isometry with fixed point A and angle of rotation ϕ is given by

$$(W - A)/(W - 1/\bar{A}) = e^{i\phi}(Z - A)/(Z - 1/\bar{A}).$$

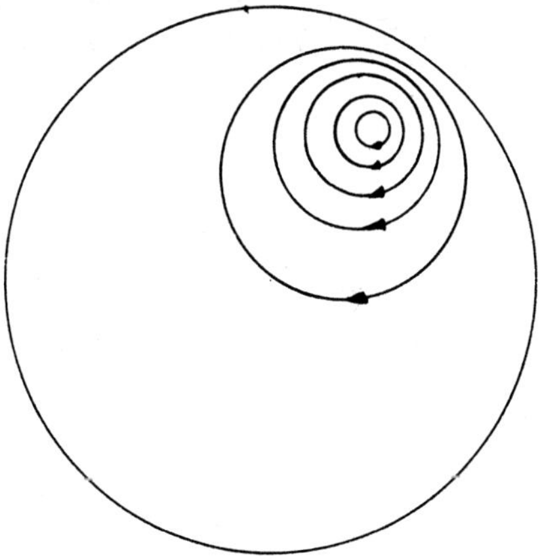

Figure 9. Orbits of an elliptic isometry.

Definition 3.7.4

Parabolic isometries, $\Delta = 1$.

These have a unique fixed point lying on the unit circle. The orbit set is the pencil of circles tangent to S^1 at the fixed point. Circles touching the circle at infinity are called *horocycles*. Again the orthogonal family is permuted by the isometry : see Figure 10.

A parabolic isometry is the 'limit' of an elliptic isometry as the internal fixed point tends to infinity.

A parabolic isometry with fixed point A is given by

$$iA/(W-A) = iA(Z-A) - k \ ,$$

where k is an arbitrary non zero real number.

Definition 3.7.5

Hyperbolic isometries or translations, $\Delta > 1$.

This case is dual to the elliptic case. Hyperbolic isometries have two fixed points lying on the unit circle. The orbit set is

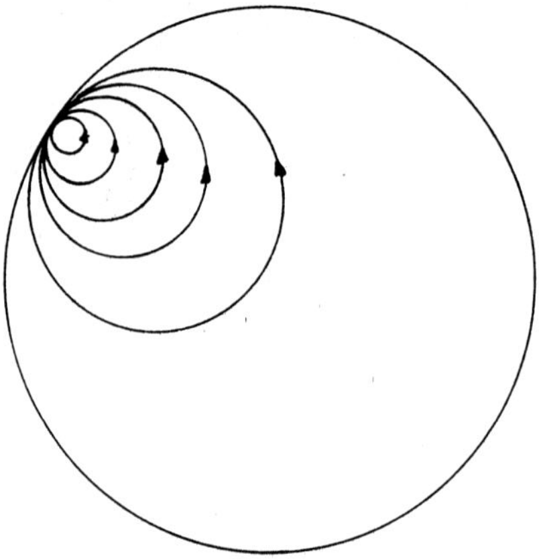

Figure 10. Orbits of a parabolic isometry.

the pencil of circles (and the one straight line) passing through these fixed points. One of these circles defines an H - line called the *axis* of the isometry. The other circles are sometimes called *hypercycles* or equidistant curves. They are the set of points 'equidistant' from the axis.

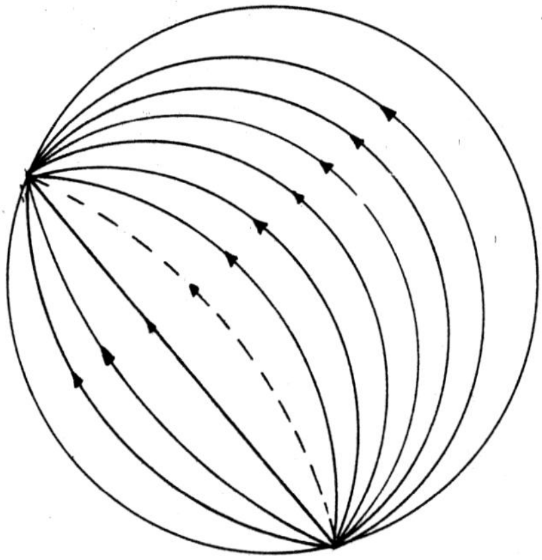

Figure 11. Orbits of a hyperbolic isometry. The axis is dotted.

A hyperbolic isometry can be thought of as a translation along the axis through a distance $2\ cosh^{-1}\sqrt{\Delta}$.

One fixed point of a hyperbolic isometry is a source and the other a sink. This allows the following useful characterisation of hyperbolic isometries: suppose that an isometry h takes the oriented line L_1 into the oriented line L_2. The lines L_1 and L_2 are not to meet even at infinity, i.e. they are ultra-parallel. Let H_1 be the part of H to the right of L_1. Define H_2 similarly. If $H_1 \subset H_2$ or $H_2 \subset H_1$ then h is hyperbolic. In the first case the source lies in H_1, in the second case the source lies in H_2.

Another useful characterisation is that for all other types of isometry h, there are points Z and hZ arbitrarily close to one another.

A hyperbolic isometry with fixed points A and B is given by

$$(W-A)/(W-B) = k(Z-A)/(Z-B),$$

for some real k.

3.7.6 The construction of the fundamental polygon P

To help understand the following the reader should consult Figure 12, which illustrates the case $n=8$. Let $X_1, X_2, \ldots, X_n$ be the n-th roots of unity where $n = 4\gamma$, $\gamma > 1$. Let $Y_1, Y_2, \ldots, Y_n$ be points on S^1 so that Y_i is midway between X_i and X_{i+1}. (Here i is a cyclic index taking the values $i = 1, 2, \ldots, n, \bmod n$.) Consider the H-lines $X_i X_{i+1}$. These form a regular n-gon with vertices at infinity and internal angle equal to zero. The half spaces of the H-lines $Y_i Y_{i+2}$ determine a regular compact n-gon with internal angle equal to $2\alpha'$ say. The regular n-gon P with internal angle equal to 2α is intermediate between the others. Note that 2α can take any value in the range $0 < 2\alpha < 2\alpha'$.

By considering the triangle OAB it can be seen that

$$2\alpha' + 2\pi/n < \pi \quad \ldots \quad 1.$$

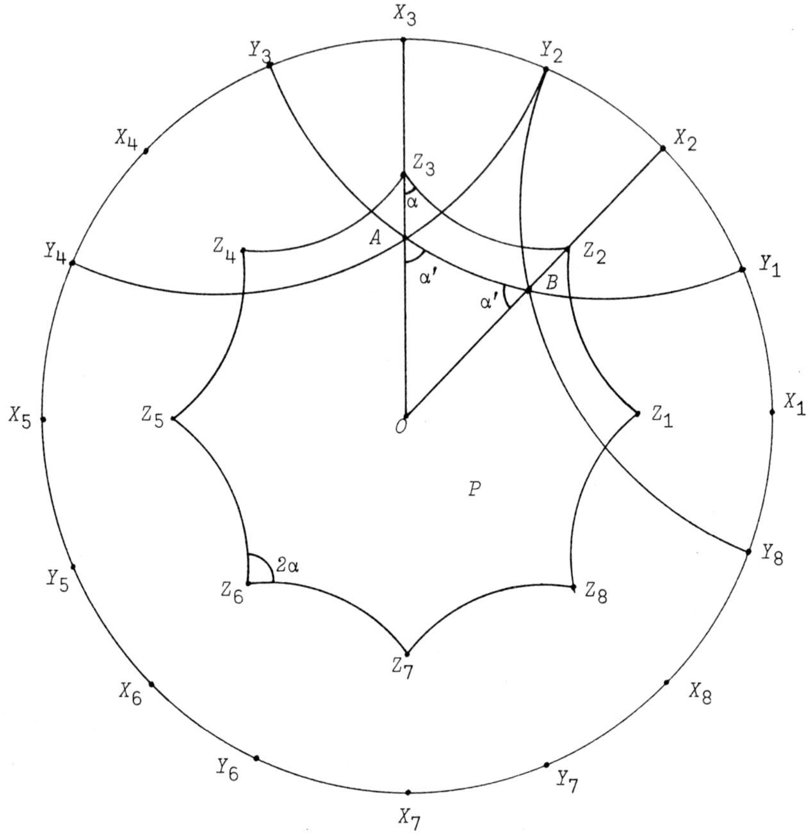

Figure 12. The fundamental polygon for the case $n = 8$.
(Not drawn to scale.)

By considering the triangle AY_2B it can be seen that

$2\pi - 4\alpha' < \pi$ ... *2* .

The inequalities *1* and *2* imply

$\pi/4 < 2\alpha' < \pi(1 - 2/n)$.

By continuity P may be chosen so that its internal angle $2\alpha = 2\pi/n$.

3.7.7 The hyperbolic isometries a_i and b_i.

Let the vertices of P be $Z_1, Z_2, \ldots, Z_n$ in anti-clockwise cyclic order. Let a_1 be the isometry which takes Z_3Z_4 to Z_2Z_1 and let b_1 take Z_2Z_3 to Z_4Z_5 : see Figure 13.

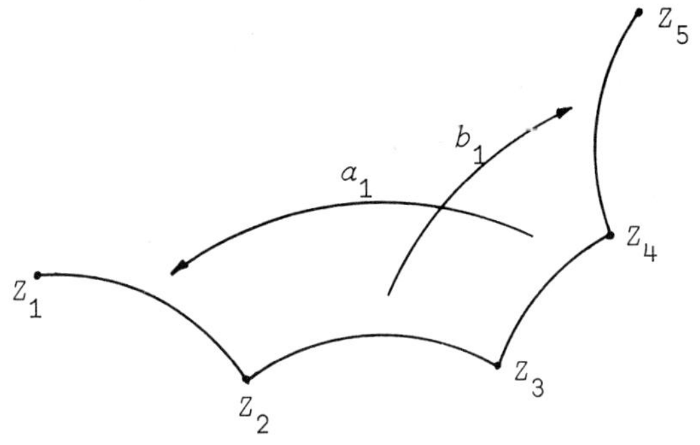

Figure 13. The isometries a and b.

Looking at Figure 12 it can be seen that the half spaces defined by the H - lines through Z_3Z_4 and through Z_2Z_1 are nested. Therefore a_1 (and similarly b_1) is hyperbolic. However, this will also follow later as part of a more general result.

A quick calculation shows that if $c_1 = a_1b_1a_1^{-1}b_1^{-1} = [a_1, b_1]$ then $c_1Z_5 = Z_1$. Let a_i, b_i and c_i be defined cyclically, $i = 1,2,\ldots,\gamma$. Then $c_iZ_{4i+1} = Z_{4i-3}$, $i = 1,2,\ldots,\gamma$.

Consider the group G of all isometries generated by $a_1, b_1, a_2, b_2, \ldots, a_\gamma, b_\gamma$. If g is an element of G let $gP = g(P)$ be the image of P under the isometry g. The element $c_1c_2\ldots c_\gamma$ of G keeps the vertices P fixed and is therefore the identity.

As g varies over the 4γ elements
$g = a_1,\ a_1b_1,\ a_1b_1a_1^{-1}, \ldots,\ c_1c_2\ldots c_{\gamma-1}a_\gamma b_\gamma a_\gamma^{-1}$,

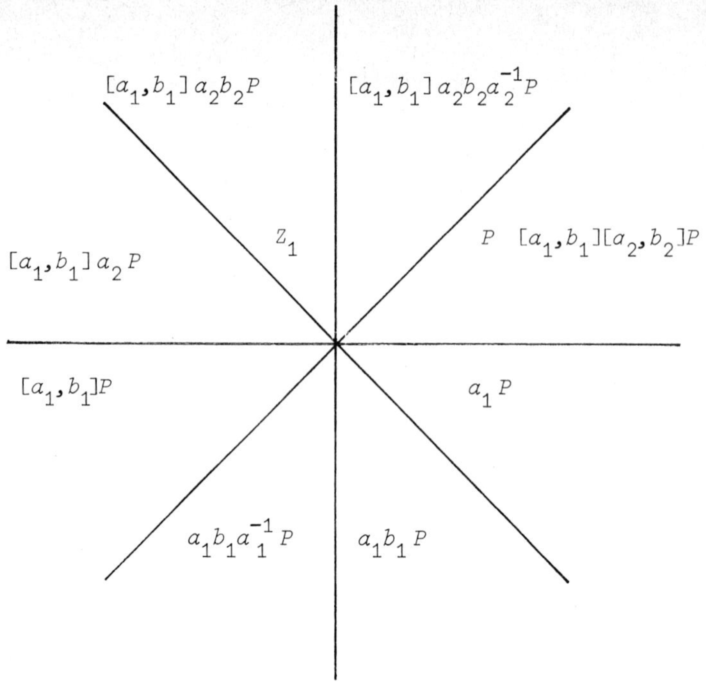

Figure 14. The polygons all fit round a vertex without overlapping.

$c_1c_2 \ldots c_{\gamma-1}c_\gamma = 1$, the polygons gP share the vertex Z_1 and the corresponding angles are lined up clockwise around Z_1 : see Figure 14 for the case $\gamma = 2$. The interiors of adjacent polygons are disjoint for the following reasons: Consider, say, the intersection of a_1b_1P and $a_1b_1a_1^{-1}P$. This is a translation of the intersection of P and $a_1^{-1}P$ which only consists of their common edge. Since P was chosen to have internal angle equal to $2\pi/4\gamma$ the 4γ polygons gP form a full neighbourhood of Z_1. By a cyclic permutation of the 4γ factors in $c_1c_2 \ldots c_\gamma$ the same is true for every other vertex of P. Furthermore, if gZ_i is a vertex of gP the polygons which form a neighbourhood of Z_i are transformed to form a neighbourhood of gZ_i.

3.7.8 The tiling of the plane by the polygons gP

In this section it will be shown that the translates gP of the fundamental polygon P cover the whole plane without overlapping. This tiling is discussed in the books by Magnus and by Siegel. It has already been shown that the polygons fit nicely along the edges and about the vertices. It remains to show that the polygons do not overlap globally. For the case $\gamma = 2$ see Figure 15. The idea behind the proof is to consider the space $\tilde{H}$ obtained from the union of the polygons by not allowing them to overlap at a distance and then to show that $\tilde{H}$ actually is H!

More specifically, let $g\tilde{P}$ be a copy of the polygon gP for each g in G. Assume initially that the $g\tilde{P}$ are mutually disjoint. Let X be the set of generators of G and let X^{-1} be the set of inverses. So

$$X = \{a_1, a_2, \ldots, a_\gamma, b_1, b_2, \ldots, b_\gamma\}$$

$$X^{-1} = \{a_1^{-1}, a_2^{-1}, \ldots, a_\gamma^{-1}, b_1^{-1}, b_2^{-1}, \ldots, b_\gamma^{-1}\}.$$

Then $g\tilde{P}$ and $h\tilde{P}$ are to be glued along an edge if and only if $gh^{-1} \in X \cup X^{-1}$. In this case gP and hP will have an edge in common in H. This common edge will define the glueing of $g\tilde{P}$ and $h\tilde{P}$. The two end points of the edge are also to be identified in the obvious way. Let $\tilde{H}$ denote the resulting space. So $\tilde{H} = \cup g\tilde{P}$ with the above identifications. Notice that $\tilde{H}$ has the same local structure as H. There is a map $p : \tilde{H} \to H$ which takes $\tilde{P}$ identically to its copy P. Locally distances, angles and indeed geometry are identical and are identified by the map p. The polygons $g\tilde{P}$ cover $\tilde{H}$ completely without overlapping.

Firstly let us show that p is onto. Because H is connected it is only necessary to show that $\cup\, gP$ is both open and closed in H. It is not hard to see that $\cup\, gP$ is open. This is because if the point B lies in gP it either lies in its interior, or in the interior of an edge of gP which is also the edge of an adjacent hP, $gh^{-1} \in X \cup X^{-1}$, or it is a vertex of gP in which case it is surrounded by an open neighbourhood of

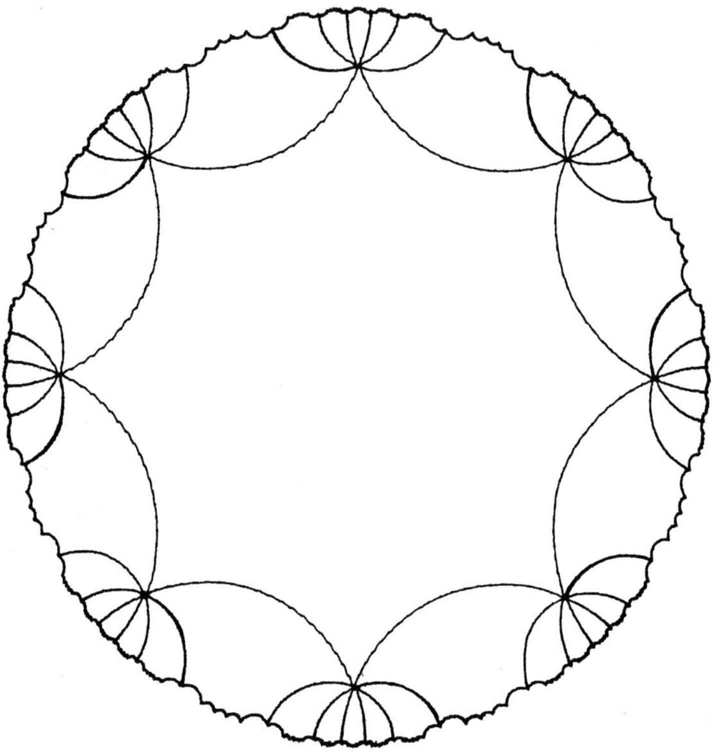

Figure 15. The network of polygons when $\gamma = 2$.

translates of P. So let A be a point of H which does not lie in $\cup gP$ but does lie in its closure $\overline{\cup gP}$. It will be shown that points such as these do not exist.

Every neighbourhood of A must meet at least one gP, in fact an infinite number of them. Let us suppose that this neighbourhood is the disc of radius ε and centre A where ε is to be a suitably chosen small number.

Now there is a positive number δ such that if B lies in gP and C is a distance less than δ from B then either C lies in gP or C lies in hP where $gh^{-1} \in X \cup X^{-1}$. A suitable choice for δ could be one half the length of any side of P. But this leads to a contradiction if ε is put equal to δ because A was supposed not to lie in $\cup gP$.

Therefore the closure of $\cup gP$ is the whole of the hyperbolic plane and so the map p is onto H.

It will now be shown that p is a covering map and because H is simply connected is 1-1. But p clearly has the path lifting property of 3.2.11 and satisfies the hypotheses of 3.2.12. So p is a covering map and therefore the polygons gP cover H completely without overlapping.

3.7.9 The orbit space as T_γ.

We have seen that there is a $1-1$ correspondence between G and the polygons gP. The elements of G are isometries which act properly discontinuously so that the orbit space H/G is covered by H. The orbit space can be obtained from just one polygon, say P, by identifying edges $e_1 = Z_i Z_{i+1}$ with $e_2 = Z_j Z_{j+1}$ if there is a group element transporting e_1 into e_2. Now to any edge e there is a unique disjoint edge f and a unique generator or its inverse in $X \cup X^{-1}$ which maps e to f. Conversely, associated with each generator there is a unique pair of edges which are identified by this generator. The result of identifying P by these generators is the orientable surface of genus γ. See Figure 16 for the identifications when $\gamma = 2$.

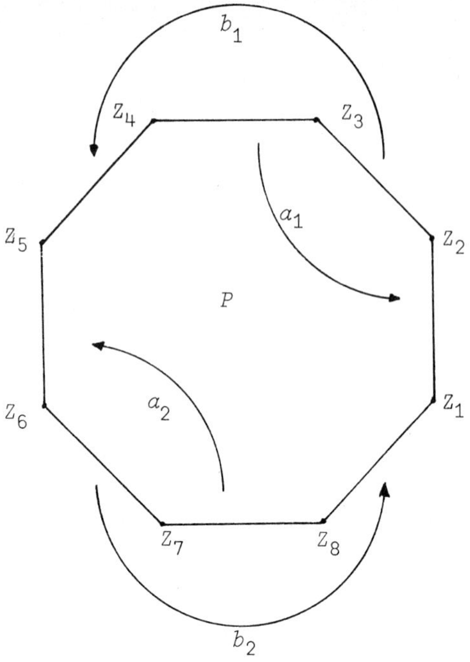

Figure 16. Identifying the polygons when $\gamma = 2$.

It only remains to show that these are the only identifications.

Let e and f be two edges of P and g an element of G which takes e into f. Let xP be the polygon having the edge f in common with P where $x \in X \cup X^{-1}$. Then the element $x^{-1}g$ keeps P fixed and is therefore the identity. So the only identifications are the 2γ ones given by the generators.

Lemma 3.7.10. *All elements of* $G - \{1\}$ *are hyperbolic.*

Proof. Suppose that g is a non-trivial element of G which is not hyperbolic. Then there would be points Z, gZ in H arbitrarily close to one another. However, there is a positive minimum for the distances between any two points in P which are to be identified, and this is therefore a minimum for all identified pairs in H. This is a contradiction, so all elements of $G - \{1\}$ are hyperbolic.

3.8 Branched covering spaces.

Consider the complex function $f(z) = z^{1/n}$ where n is a positive integer. The graph of $f(z)$, Γ, which can be used to construct the n-*fold* cover of $\mathbb{C}^* = \mathbb{C} - \{0\}$ can be extended over 0. Now 0 has just one *n th root*, namely 0, so if $\Gamma^+ = \{(z, f(z)) \mid z \in \mathbb{C}\}$ and if $p : \Gamma^+ \longrightarrow \mathbb{C}$ is the projection $p(z, z^{1/n}) = z$, then p is a covering map when restricted to $p^{-1}\mathbb{C}^* = \Gamma$ and is a covering map when restricted to $p^{-1}0 = \{(0, 0)\}$ but fails to be a covering map over the whole of Γ^+. Such a map is called a branched covering map with upstairs branching set $\{(0, 0)\}$ and downstairs branching set $\{0\}$. A general definition will be given shortly. In the meantime, note that $z^{1/n}$ can be extended to the Riemann sphere by defining $\infty^{1/n} = \infty$. This extended mapping may be described as follows: Take a sphere, thinking of it as the surface of the earth and cut it down from the North Pole to the South Pole along n regularly spaced lines of longitude. Each two-sided spherical region is stretched around the whole world until its longitudinal edges come to rest along a fixed longitude, say the Greenwich meridian. The North Pole corresponding to $z = \infty$ and the South Pole corresponding to $z = 0$ stay fixed. In analytic terms the branched cover is given by $z \longrightarrow z^n$, with 0 and ∞ as branch points.

This example can be generalised to k dimensions as follows: Consider $D^2 \times I^{k-2}$, homeomorphic to the k-ball D^k where D^k is the set of points with distance not greater than one from the origin and I is the closed interval $[0, 1]$.

Let $f_n : D^2 \times I^{k-2} \longrightarrow D^2 \times I^{k-2}$ be defined by $f_n(z, t_1, \ldots, t_{k-2}) = (z^n, t_1, \ldots, t_{k-2})$. Here the branching set both upstairs and downstairs is $\{0\} \times I^{k-2}$.

Using this example as a local condition, the following definition can be made:

Definition 3.8.1

A map $p : \tilde{X} \longrightarrow X$ between metric spaces is a **branched cover** if every point has a neighbourhood which looks like the above example. That is, if x lies in X then x has a neighbourhood B such that if $\tilde{B}$ is a component of $p^{-1}B$ then there is a commuting diagram

$$\begin{array}{ccc} D^2 \times I^{k-2} & \xrightarrow{\;\tilde{h}\;} & \tilde{B} \\ {\scriptstyle f_n}\downarrow & & \downarrow{\scriptstyle p\,|\,\tilde{B}} \\ D^2 \times I^{k-2} & \xrightarrow{\;h\;} & B \end{array}$$

where h and $\tilde{h}$ are homeomorphisms with $h(0, \tfrac{1}{2}, \ldots, \tfrac{1}{2}) = x$, [13]. The integer n is called the **branching index** of $\tilde{x} = p^{-1}x \cap \tilde{B}$. The set of points $\tilde{\Sigma}$ with branching index > 1 is called the **upstairs branching set.**

The image $\Sigma = p(\tilde{\Sigma})$ is the **downstairs branching set.**

Several consequences follow immediately from the definition.

1. X and $\tilde{X}$ are both k dimensional manifolds.
2. Σ and $\tilde{\Sigma}$ are both codimension 2 submanifolds.
3. $p \,|\, \tilde{X} - \tilde{\Sigma}$ and $p \,|\, \tilde{\Sigma}$ are both unbranched covers, [14].
4. The branching index is locally constant on $\tilde{\Sigma}$ and so is constant on any component.

Definition 3.8.2

A branched cover $\tilde{X} \longrightarrow X$ is said to be a p-**fold (regular) cover** if the unbranched cover $\tilde{X} - \tilde{\Sigma} \to X - \Sigma$ is a p-*fold* (regular) cover.

Theorem 3.8.3 *Every closed orientable surface* T_γ *is a* 2-*fold regular branched cover of the two sphere* T_0.

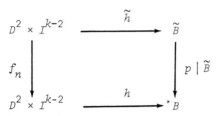

Proof. Actually rather more will be proved. The surface T_γ is the boundary of the solid ball with γ handles H_γ. Let $C_1, \ldots, C_\gamma$ be a chain of circles of equal radius *1* lying in the plane $y = 0$ with centres at the points $(1,0,0), (3,0,0), \ldots, (2\gamma-1, 0, 0)$, see *figure 17*.

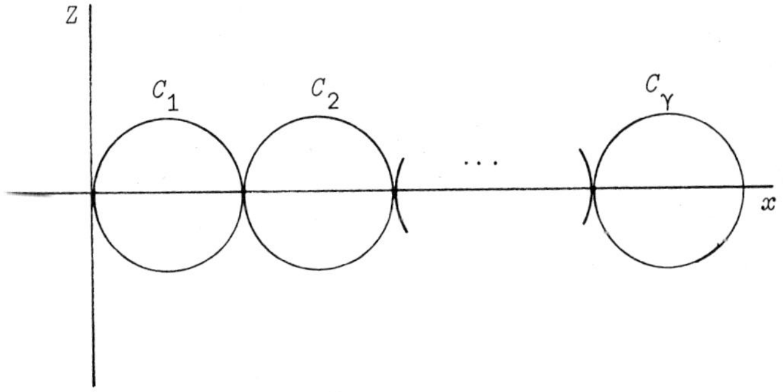

Figure 17.

The space H_γ can be thought of as a 'thickening' of $\bigcup_{i=1}^{\gamma} C_i$.

Let H_γ^+ be those points of H_γ with non-negative 2 - *coordinate*, and H_γ^- those with non-positive 2 - *coordinate*. It is not hard to see that each $H_\gamma^\pm$ is topologically a ball. Their intersection $H_\gamma^+ \cap H_\gamma^-$ is the intersection of H_γ with the plane $z = 0$ and consists of the disjoint union of $\gamma + 1$ discs with centres at $(0,0,0), (2,0,0), \ldots, (2\gamma,0,0)$. Let $\tau : R^3 \longrightarrow R^3$ be reflection in the *x-axis*, $\tau(x,y,z) = (x, -y, -z)$. Assume that $\tau H_\gamma^+ = H_\gamma^-$ and that $\tau H_\gamma = H_\gamma$; i.e. H_γ is symmetric enough to be invariant under the map τ. Consider the space B obtained from H_γ by identifying points P with τP. Then B is obtained from H_γ^- by identifying P with τP on $H_\gamma^+ \cap H_\gamma^-$ and is clearly homeomorphic to a ball.

If $p : H_\gamma \longrightarrow B$ is the natural map, then p is a two-fold regular cover branched over those parts of the *x-axis* which lie in H_γ.

The restriction $p \mid T_\gamma : T_\gamma \longrightarrow \partial B$ is the required map.

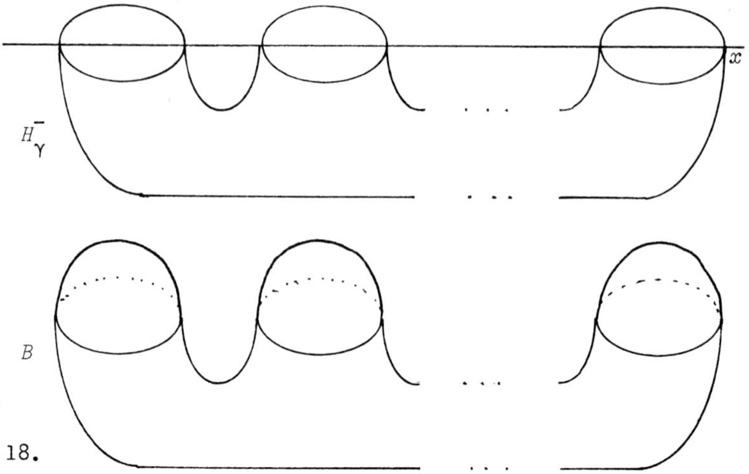

Figure 18.

This theorem cannot be generalised to higher dimensions. For instance, $S^1 \times S^1 \times S^1$ is not the two fold branched cover of a three sphere. However, the following theorem was proved by Alexander in 1919 !

Theorem 3.8.4 *Any compact closed orientable 3-manifold is the branched cover of the three sphere.*

Proof. Let the manifold be triangulated by the complex K, [15] . Let the vertices of K be $v_1, \ldots, v_n$. Pick n distinct points $t_1, \ldots, t_n$ on the curve $C = \{(t, t^2, t^3) \mid t \in R^3\}$ $\subset R^3 \cup \{\infty\} = S^3$. Points on the curve C have a natural ordering given by their parameter t. Four points $t_1 < t_2 < t_3 < t_4$ define a bounded tetrahedron $In(t_1, t_2, t_3, t_4)$ with vertices defined by these parameters. This tetrahedron is non-degenerate because the determinant

$$\begin{vmatrix} 1 & t_1 & t_1^2 & t_1^3 \\ 1 & t_2 & t_2^2 & t_2^3 \\ 1 & t_3 & t_3^2 & t_3^3 \\ 1 & t_4 & t_4^2 & t_4^3 \end{vmatrix} = \prod_{j<k} (t_j - t_k) \neq 0 .$$

The vertices also define a complementary tetrahedron $Out\,(t_1, t_2, t_3, t_4)$ containing ∞.

There is an obvious map from the *2 - skeleton* $f : K^2 \longrightarrow S^3$ given by $g(v_i) = t_i$, $i = 1,\dots, n$. However, for each tetrahedron in K there is a choice between the inner and outer corresponding tetrahedron in S^3. By hypothesis the tetrahedra of K can be oriented coherently so that their algebraic sum forms a *3 - cycle*. Consider a tetrahedron T in K. Its boundary ∂T has an induced orientation as does $f(\partial T)$ in S^3, [16]. The oriented sphere $f(\partial T)$ is the oriented boundary of either the inside or the outside, provided S^3 has some fixed orientation. Define $f(T)$ so that $\partial f(T) = f(\partial T)$.

Clearly $f : K \longrightarrow S^3$ is a covering map on the interior of each tetrahedron.

Let x be a point in the interior of a *2 - simplex* τ of K. Suppose that τ is a face of T_1 and T_2. By the orientability condition $f(T_1)$ and $f(T_2)$ lie on opposite sides of $f(\tau)$. So f is still an unbranched cover when restricted to the interior of each *2 - simplex*.

Let E be a *1 - simplex* of K. There is a simple polygonal loop $C = v_{i_1}, v_{i_2}, \dots, v_{i_k}$, the link of E in K, such that the tetrahedra $(E, v_{i_j}, v_{i_{j+1}})$ lie in K, $j = 1, 2,\dots, k$. Two things may happen:

1. If C_j is the arc complementary in C to the edge $(v_{i_j}, v_{i_{j+1}})$, and if fC_j lies outside $f\,(E, v_{i_j}, v_{i_{j+1}})$ for some j then this is true for all $j = 1, 2,\dots, k$.
2. If fC_j lies inside $f\,(E, v_{i_j}, v_{i_{j+1}})$ for some j then this is also true for all $j = 1, 2,\dots, k$.

Properties 1 and 2 are a consequence of the orientability condition.

If 1. holds then f restricted to all interior points of E is an unbranched cover.

If 2. holds then f is a branched cover along E with

branching index $k-1$.

Finally, consider f restricted to a vertex v of K. Let B be a small neighbourhood of v such that $f \mid \partial B : \partial B \longrightarrow f(\partial B)$ is a branched cover of the two sphere by the two sphere. Let there be λ branch points downstairs. Above each downstairs branch point is a single branch point corresponding to an edge of K crossing ∂B. Suppose that f is a μ-*fold* covering away from the branch points. Then counting the Euler number for the sphere gives the formula $2 = \lambda + \mu(2-\lambda)$. So either $\mu = 1$ or $\lambda = 2$. In the first case the covering is unbranched at v. In the second case the vertex lies on part of an arc which has branching index μ.

The following more exact theorem is due to Montesinos and Hilden.

Theorem 3.8.5 *Every closed connected orientable 3-manifold is a 3-fold, possibly irregular, branched cover of* S^3 *with downstairs branch set a single simple closed curve (a knot).*

Proof. See references.

This section is concluded with an example of an irregular branched cover of the complex plane.

An example.

Let $\mathbb{C}^{**} = C - \{\pm 1\}$ with base point 0. Then $\pi_1(\mathbb{C}^{**})$ is the free group with generators a and b, see Figure 19.

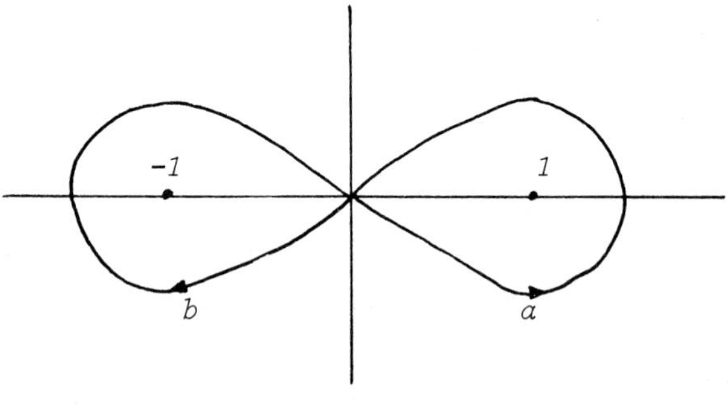

Figure 19.

Let S_3 be the symmetry group on $\{1, 2, 3\}$ with six elements. Define the transitive map $\pi_1(\mathbb{C}^{**}) \xrightarrow{m} S_3$ by $m(a) = (1, 2)$ and $m(b) = (2, 3)$. To construct the cover with corresponding monodromy make cuts along the x-axis $x \geq 1$ and $x \leq -1$. Consider three copies X_1, X_2, X_3 of the cut plane with identifications as shown.

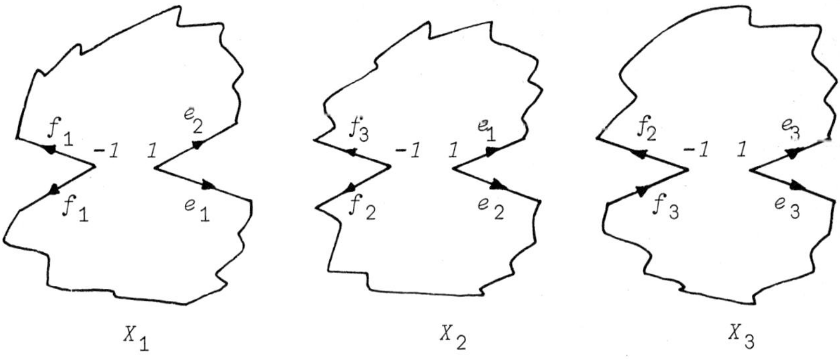

Figure 20.

These are glued up in such a way that on traversing a for example the level X_2 is reached from X_1 and X_3 stays fixed. This leads to the glueing pattern described in Figure 20. The result is another copy of $\mathbb{C}$.

Notes on Chapter 3.

1. See Ahlfors & Sario.

2. Since $\log z = \log |z| + i \arg z$ the graph of $\log z$ is similar.

3. Unlike $\arg z$ and $\log z$, $z^{\frac{1}{2}}$ is defined when $z = 0$. This is important when considering branched covers later.

4. This homeomorphism can be required to preserve special properties according to circumstances. Put another way, structure in U can be lifted to $\tilde{U}$. This is the case for example with Riemann surfaces which are locally like the complex plane.

5. The base point $\tilde{x}_0$ is important; see the example after lemma 3.6.8.

6. The local condition on Y is necessary, see Hilton & Wylie pp. 258-259 .

7. This is the same definition as that given in Chevalley. Massey calls a space semi locally simply connected and Hilton & Wylie call a space locally simply connected in the weak sense if every point has a neighbourhood in which every loop contracts in the ambient space.

8. Called a glide reflection.

9. All spaces are assumed locally connected and locally simply connected.

10. Not to be confused with a one-humped camel.

11. Permutations always act on the right to preserve the group operation.

12. To see this, count Euler numbers.

13. A more general definition can be found in Fox's paper on branched covers.

14. Possibly disconnected covers.

15. See Moïse.

16. Another way of looking at this is to note that an orientation of a 2- simplex defines an outward normal if the ambient space is oriented.

4. THE HOMOLOGY OF COVERING SPACES

'Whoever has found his path and follows it,
keeping straight ahead and climbing up,
fulfills himself.'

Reinhold Messner *Solo Nanga Parbat*

4.1 Reidemeister Chains

Let K be a connected cell complex. By the previous chapter K has a universal cover $\tilde{K}$. Each cell in K can be lifted to provide a cell decomposition of $\tilde{K}$. Assume that K has just one vertex e which we will take as a base point and choose once and for all a base point $\tilde{e}$ in $\tilde{K}$ above e.

If $\phi : D^p \rightarrow K$ is the characteristic map of a p-cell σ in K with $\phi(1,0,\ldots,0) = e$ there is a unique lift $\tilde{\phi} : D^p \rightarrow \tilde{K}$ with $\tilde{\phi}(1,0,\ldots,0) = \tilde{e}$ which is characteristic for the p-cell $\tilde{\sigma}$ in $\tilde{K}$. Any other lift of σ is of the form $g\tilde{\sigma}$ where g lies in $\pi = \pi_1(K, e)$.

The chains in $C_p(\tilde{K}; \mathbb{Z})$ can therefore be written in the form $\sum \lambda_i \tilde{\sigma}_i$ where σ_i varies over the p-cells of K and λ_i lies in the integer group ring $\mathbb{Z}\pi$ [1].

Symbolically, this may be written

$$C_p(\tilde{K}) \approx \mathbb{Z}\pi \otimes_{\mathbb{Z}} C_p(K), \quad [2].$$

To complete the description of $\tilde{K}$ it is necessary to describe the boundary homomorphisms $\partial : C_p(\tilde{K}) \rightarrow C_{p-1}(\tilde{K})$. We shall do this for the cases $p \leq 3$.

By linearity the homomorphisms ∂ will be determined once their values on the lifted cells $\tilde{\sigma}$ are known. If a cell $\tilde{\sigma}$ has an identification on its boundary which creates elements of the fundamental group then these identifications unfold in the universal cover.

The following conventions will be adopted for the fundamental group. As before, the elements $a, b, c, \ldots$ etc. of a set A will denote both the 1-cells of K and the generators of the free group $F = \pi_1(K^1, e)$. The image of these generators in $\pi = \pi_1(K, e)$ will be denoted by the corresponding Greek letters $\alpha, \beta, \gamma, \ldots$ etc.

p = 1 Let $\tilde{a}$ be the lift of the 1-cell a with initial point $\tilde{e}$. Then the boundary is given by

$$\partial \tilde{a} = \alpha \tilde{e} - \tilde{e} = (\alpha - 1)\tilde{e} .$$

p = 2 If $\tilde{r}$ is the lift of a 2-cell r with boundary word $w(r)$ then $\partial \tilde{r} = \sum_{a \in A} (\partial_a w)\tilde{a}$ for some elements $\partial_a w$ in the group ring $\mathbb{Z}\pi$. The operators $\partial_a w$ are the image in $\mathbb{Z}\pi$ of operators $d_a : \mathbb{Z}F \longrightarrow \mathbb{Z}F$ satisfying:

1. $$d_a(b) = \begin{cases} 1 & \text{if } a = b \\ 0 & \text{if } a \neq b \end{cases}$$

2. $$d_a(ts) = d_a(t) + t\,d_a(s), \quad ts = w .$$

The reason for rule 2 can be deduced from Figure 1.

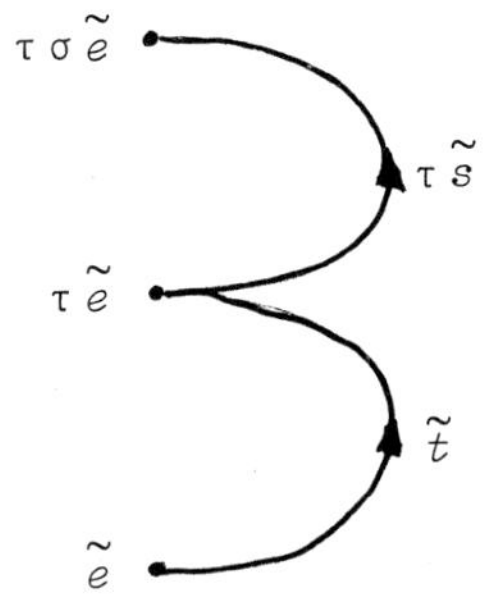

Figure 1.

These two rules determine the values of d_a and hence ∂_α.

Example. Let $w = a b a^{-1} b$ then

$$d_a w = 1 - a b a^{-1} \qquad d_b w = a + a b a^{-1}.$$

We will have more to say about d_a and ∂_a shortly.

p = 3 Any 3-cell x in K is attached to K^2 by a map $f : S^2 \to K^2$. If f is transverse to K^2 then we have seen in Chapter 2 how f is determined by an identity of the form

$$(f_1 \cdot r_1)(f_2 \cdot r_2) \ldots (f_k \cdot r_k) \equiv 1 \quad f_i \in F, \ r_i \in R.$$

Its boundary may be specified by the element $\phi_1 \tilde{r}_1 + \ldots + \phi_k \tilde{r}_k$ in the kernel of the map $C_2(\tilde{K}) \xrightarrow{\partial} C_1(\tilde{K})$. By the Roman to Greek convention ϕ_i is the element of π determined by the element f_i in F.

Example. Consider the following cell decomposition K of the 3-torus $S^1 \times S^1 \times S^1$. The 2-skeleton of K is

$$K^2 = \{a, b, c \mid r = [a, b], \ s = [b, c], \ t = [c, a]\}$$

and K has one 3-cell attached by the identity

$$(a \cdot s^{-1})(1 \cdot r)(b \cdot t^{-1})(1 \cdot s)(c \cdot r^{-1})(1 \cdot t) \equiv 1.$$

The boundary of the 1-cells are given by

$$\tilde{a} \to (\alpha - 1)\tilde{e}, \quad \tilde{b} \to (\beta - 1)\tilde{e}, \quad \tilde{c} \to (\gamma - 1)\tilde{e}.$$

The boundary of the 2-cells are given by

$$\begin{aligned}
\tilde{r} &\to (1-\beta)\tilde{a} - (1-\alpha)\tilde{b} \\
\tilde{s} &\to (1-\gamma)\tilde{b} - (1-\beta)\tilde{c} \\
\tilde{t} &\to -(1-\gamma)\tilde{a} + (1-\alpha)\tilde{c}.
\end{aligned}$$

The boundary of the 3-cell is

$$(1-\alpha)\tilde{s} + (1-\beta)\tilde{t} + (1-\gamma)\tilde{r} .$$

These boundary maps are summarised in Figure 2.

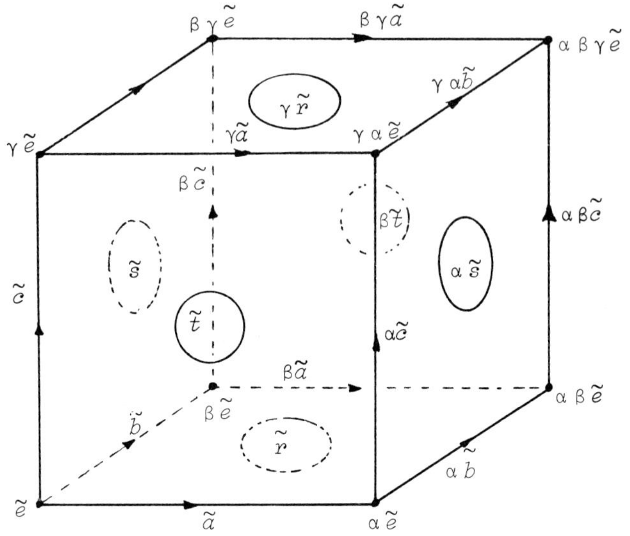

Figure 2. The boundary of the fundamental cube in $S^1 \times S^1 \times S^1$.

4.2 Properties of the Fox derivatives.

The formula for the boundary of a lifted 2-cell justifies the following:

Definition 4.2.1 Derivation

If π is any group let $\varepsilon : \mathbb{Z}\pi \longrightarrow \mathbb{Z}$ be the linear map, called **augmentation** which takes group elements into 1.

A map $d : \mathbb{Z}\pi \longrightarrow \mathbb{Z}\pi$ is called a **derivation** if

1. d is linear with respect to addition, so $d(\lambda + \mu) = d(\lambda) + d(\mu)$.
2. $d(\lambda\mu) = d(\lambda)\varepsilon(\mu) + \lambda d(\mu)$.

Note that several consequences of 1. and 2. follow immediately, namely:

$$d(n\lambda) = n\,d(\lambda)$$

$$d(n) = 0$$

$$d(g^{-1}) = -g^{-1}d(g) \quad \text{for} \quad n \in \mathbb{Z} \quad g \in \pi .$$

For group elements condition 2. takes the slightly simpler but equivalent form

2' $\quad d(gh) = d(g) + g\,d(h)$.

An example of a derivation is the map D given by

$$D(\lambda) = \lambda - \varepsilon(\lambda) .$$

The set of derivations forms a right $\mathbb{Z}\pi$ *-module* where addition is defined by $(d_1 + d_2)(\lambda) = d_1(\lambda) + d_2(\lambda)$ and right multiplication is defined by $(d\mu)(\lambda) = d(\lambda)\mu$. For the following lemma we return to the free group F .

Lemma 4.2.2

To each generator a *of* F *there corresponds a derivation* d_a, *called derivation with respect to* a, *which has the property that*

$$d_a(b) = \begin{cases} 1 & \text{if} \quad a = b \\ 0 & \text{if} \quad a \neq b . \end{cases}$$

Furthermore, if d *is a derivation satisfying*

$$d(a) = \lambda_a , \quad \textit{for certain elements} \quad \lambda_a \in \mathbb{Z}F , \ a \in A ,$$

then $d(\lambda) = \sum d_a(\lambda)\,\lambda_a$ *for all* $\lambda \in \Lambda$.

i.e. $d = \sum d_a \lambda_a$.

Proof. The existence of the derivatives d_a follows from the following formula:

Let $w = w_1 a^{\varepsilon_1} w_2 a^{\varepsilon_2} \ldots w_k a^{\varepsilon_k} w_{k+1}$ where $\varepsilon_i = \pm 1$ and $w_1, \ldots, w_{k+1}$ are words in F which do not involve a, then

$$d_a w = \sum_{i\ 1}^{k} \varepsilon_i w_1 a^{\varepsilon_1} w_2 a^{\varepsilon_2} \ldots w_{i-1} a^{\varepsilon_{i-1}} w_i a^{(\varepsilon_i - 1)/2} .$$

The uniqueness of d_a follows from the observation that the value of any derivative is determined by its values on the basis elements.

Consider the derivative $d - \sum d_a \lambda_a$. This has value zero on the generators $a \in A$ and so equals zero.

Corollary 4.2.3 **Taylor's theorem for Fox derivatives.**

$$\lambda = \varepsilon(\lambda) + \sum (d_a \lambda)(a - 1) .$$

Proof. This is a consequence of 4.2.2 and the fact that $D = 1 - \varepsilon$ is a derivative with values $a - 1$ on the generators $a \in A$.

4.2.4 Higher order derivatives

The higher order derivatives are defined inductively by

$$d_{a_1 a_2 \ldots a_k}(\lambda) = d_{a_1}(d_{a_2 \ldots a_k})(\lambda) .$$

The integer k is called the **order** of the derivative. The operators $\varepsilon_{a_1 a_2 \ldots a_k}(\lambda) = \varepsilon d_{a\ a\ \ldots a_k}(\lambda)$ are called augmented derivatives. We have already met ε_a and ε_{ab} in Chapter 1. The following lemma, which is easily proved by induction, is useful for calculating values of $\varepsilon_{a_1 \ldots a_k}$.

Lemma 4.2.5 *If $g \in F$ and $a_1 a_2 \ldots a_k$ satisfies $a_i \neq a_{i+1}$, $i = 1, \ldots, k-1$, then $\varepsilon_{a_1 a_2 \ldots a_k}(g)$ is the total number of signed occurrences [3] of $a_1 a_2 \ldots a_k$ in the word g.*

Definition 4.2.6

We will call $I = a_1 a_2 \dots a_k$ a **string** in A of **length** $k = \ell(I)$. Two strings I_1, I_2 of length ℓ_1, ℓ_2 respectively, can be juxtaposed to form the product string $I_1 I_2$ of length $\ell_1 + \ell_2$.

Lemma 4.2.7

For any string I *and* $\lambda, \mu \in \mathbb{Z}F$

$$\varepsilon_I(\lambda\mu) = \sum_{I_1 I_2 = I} \varepsilon_{I_1}(\lambda)\, \varepsilon_{I_2}(\mu),$$

where the summation is taken over all ordered pairs (I_1, I_2), *such that* $I_1 I_2 = I$ *including* (I, ϕ) *and* (ϕ, I).

Proof. This is obviously true if $\ell(I) = 0$. The proof following routinely by induction on $\ell(I)$.

Corollary 4.2.7

$$\varepsilon_I(\lambda_1 \lambda_2 \dots \lambda_k) = \sum_{I_1 I_2 \dots I_k = I} \varepsilon_{I_1}(\lambda_1)\, \varepsilon_{I_2}(\lambda_2) \dots \varepsilon_{I_k}(\lambda_j).$$

With some change in the summation convention, we also have:

Corollary 4.2.8

If $I \neq \emptyset$, $g \in F$,

$$\varepsilon_I(g^{-1}) = \sum_{I_1 I_2 \dots I_k = I} (-1)^k\, \varepsilon_{I_1}(g)\, \varepsilon_{I_2}(g) \dots \varepsilon_{I_k}(g),$$

where the sum is taken over all $I_1 I_2 \dots I_k = I$ *with* $I_j \neq \emptyset$, $j = 1, \dots, k$.

Proof. Consider the expansion of $0 = \varepsilon_I(g g^{-1})$ and use induction on $\ell(I)$.

The higher order derivatives are also linear, and satisfy the following multiplicative rule:

Lemma 4.2.9

$$d_{a_1a_2\ldots a_k}(\lambda\mu) = \sum_{j=1}^{k} d_{a_1\ldots a_j}(\lambda)\,\varepsilon_{a_{j+1}\ldots a_k}(\mu) + \lambda\, d_{a_1\ldots a_k}(\mu).$$

Proof. Once again by induction on k.

By applying Taylor's theorem k times, we obtain:

$$\lambda = \varepsilon(\lambda) + \sum_{a_1} \varepsilon_{a_1}(\lambda)(a_1 - 1) + \sum_{a_1a_2} \varepsilon_{a_1a_2}(\lambda)(a_1 - 1)(a_2 - 1) + \ldots$$
$$+ \sum_{a_1\ldots a_{k-1}} \varepsilon_{a_1\ldots a_{k-1}}(\lambda)(a_1 - 1)\ldots(a_{k-1} - 1)$$
$$+ \sum_{a_1\ldots a_k} d_{a_1\ldots a_k}(\lambda)(a_1 - 1)\ldots(a_k - 1).$$

Letting k tend to infinity, we obtain an infinite Taylor series invented by Magnus. To state this precisely, let $\mathfrak{X}$ denote the ring of associative power series in the non-commuting variables t_a, $a \in A$. Then the **Magnus expansion** is the homomorphism $m : \mathbb{Z}\pi \longrightarrow \mathfrak{X}$ defined on generators by $m(a) = 1 + t_a$, $m(a^{-1}) = 1 - t_a + t_a^2 - t_a^3 + \ldots$. Using the rules for ε_I given above, we see that

$$m(\lambda) = \varepsilon(\lambda) + \sum_a \varepsilon_a(\lambda)\, t_a + \sum_{a,b} \varepsilon_{ab}(\lambda)\, t_a t_b + \ldots .$$

Definition 4.2.10 **The augmentation ideal.**

Let J be the kernel of the augmentation map $\varepsilon : \mathbb{Z}F \longrightarrow \mathbb{Z}$. This is a two-sided ideal called the **augmentation ideal.**

Since $f - g = g(g^{-1}f - 1)$, J is right generated by the elements $(g - 1)$, $g \in F$. Furthermore, since $(gf - 1) = g(f-1) + (g-1)$ and since $(f^{-1} - 1) = -f^{-1}(f-1)$ J is right generated by the elements $(a-1),\ a \in A$.

Thus $J = \text{ideal}_{\mathbb{Z}F}\{(a-1),\ a \in A\}$.

Writing $J^2 = J \cdot J$, $J^3 = J \cdot J^2$ etc. we have more generally that

$$J^k = \text{ideal}_{\mathbb{Z}F}\{(a_1-1)(a_2-1)\ldots(a_k-1),\ a_1, a_2, \ldots, a_k \in A\}.$$

For each $k \geq 1$ let

$$D_k(F) = F \cap (1+J^k) = \{f \in F \mid f-1 \in J^k\}.$$

Then each $D_k(F)$ is a normal subgroup of F, $F = D_1(F) \supset D_2(F) \supset \ldots$ and an arbitrary element of $D_k(F)$ is of the form

$$1 + \sum_{a_1\ldots a_k} \lambda_{a_1\ldots a_k}(a_1-1)\ldots(a_k-1).$$

Using the higher order derivative properties and the Taylor expansion we can identify the subgroups $D_k(F)$ as follows:

Lemma 4.2.11

Let $k \geq 2$. The element $f \in F$ lies in $D_k(F)$ if and only if all its augmented derivatives $\varepsilon_I(f)$ vanish for $1 \leq \ell(I) < k$.

In section 4.4 we shall identify $D_k(F)$ in another way.

Definition 4.2.12

The **length** $\ell(w)$ of a reduced word $w = x_1x_2\ldots x_k$, $x_i \in A \cup A^{-1}$, is defined to be $\ell(w) = k$.

Now let $\lambda = n_1w_1 + \ldots + n_rw_r$ belong to $\mathbb{Z}F$ where $n_i \neq 0$ and $w_i \neq w_j$, $i,j = 1,\ldots,r$. Then the **length** of λ is defined to be

$$\ell(\lambda) = \max_{i=1,\ldots r} \ell(w_i).$$

Lemma 4.2.13

If λ lies in J^k and is non-zero then $\ell(\lambda) \geq k/2$.

Proof. The result is easily seen to be true if $k = 1$ or 2. Assume the result inductively up to k and let $\lambda \in J^k$. Suppose $\ell(\lambda) < k/2$. We shall then show that $\lambda = 0$. Now $\lambda = \sum_{a \in A} (u_a a + v_a a^{-1})$ where u_a, v_a are elements of $\mathbb{Z}F$ with length less than $\ell(\lambda)$.

A short calculation shows that

$$\partial_{bc} \lambda = \sum_{a \in A} (\partial_{bc} u_a + \partial_{bc} v_a) + \partial_b u_c - \partial_b v_c, \quad b \neq c \quad \text{and}$$

$$\partial_c (\partial_c \lambda) c = \sum_{a \in A} (\partial_{cc} u_a + \partial_{cc} v_a + \partial_c u_a + \partial_c v_a) + \partial_c u_c + u_c - \partial_c v_c .$$

By the inductive hypothesis both expressions vanish. So $\partial_{bc} \lambda = 0$ $v \neq c$ and $\partial_{cc} \lambda = -\partial_c \lambda$.

Using Taylor's theorem gives

$$\partial_c \lambda = \varepsilon_c \lambda + \sum_{b \in A} \partial_{bc} \lambda (b - 1) .$$

Substituting the above formulae for $\partial_{bc} \lambda$ implies that $\partial_c \lambda = 0$ for all c.

Another application of the Taylor formula

$$\lambda = \varepsilon \lambda + \sum_{a \in A} \partial_a \lambda (a - 1)$$

gives $\lambda = 0$.

Theorem 4.2.14 (The Magnus map is injective).

Let $\lambda, \mu \in \mathbb{Z}F$ and suppose that $\varepsilon \lambda = \varepsilon \mu$, $\varepsilon_a \lambda = \varepsilon_a \mu$, $\varepsilon_{ab} \lambda = \varepsilon_{ab} \mu$, etc. Then $\lambda = \mu$.

Proof.

By hypothesis $\lambda - \mu \in J^k$ for all k. If $\lambda - \mu \neq 0$ then $\ell(\lambda - \mu) \geq k/2$ for all k, which is a contradiction. So $\lambda = \mu$.

Corollary 4.2.13.

$$\bigcap_{k=0}^{\infty} J^k = \{0\} \quad \text{and} \quad \bigcap_{k=0}^{\infty} D_k(F) = \{1\}.$$

4.3 The Lower Central Series.

Definition 4.3.1

A commutator in a group G is an expression of the form $[g, h] = g h g^{-1} h^{-1}$.

If A, B are subgroups of G, then $[A, B]$ is the subgroup generated by elements $[a, b]$, $a \in A$, $b \in B$.

The lower central series $G^{(1)} \supset G^{(2)} \supset G^{(3)} \supset \ldots$ is defined inductively by the rules

$$G^{(1)} = G$$

and $$G^{(k)} = [G^{(k-1)}, G].$$

The following lemma is easily proved, [4].

Lemma 4.3.2

$G^{(k)}$ is normal in G and $G^{(k-1)}$, moreover $G^{(k-1)}/G^{(k)}$ is Abelian.

We shall show that if F is free then $F^{(k-1)}/F^{(k)}$ is free Abelian and we shall find a set of base elements called basic commutators.

Definition 4.3.3 Basic Commutators.

Let A be a set, then a **basic commutator** in A is defined inductively as follows:

B1. Each basic commutator c has a **weight** $\ell(c)$ taking one of the values $1, 2, 3 \ldots$.

B2. The basic commutators of weight 1 are the elements of A . A basic commutator of weight >1 is a symbol of the form $c = [c_1, c_2]$ where c_1, c_2 are previously defined basic commutators and $\ell(c) = \ell(c_1) + \ell(c_2)$.

B3. Basic commutators are ordered so as to satisfy the following:
(i) basic commutators of the same weight are ordered arbitrarily;
(ii) if $\ell(c) > \ell(c')$ then $c > c'$.

B4. (i) If $\ell(c) > 1$ and $c = [c_1, c_2]$ then $c_1 < c_2$;
(ii) if $\ell(c) > 2$ and $c = [c_1, [c_2, c_3]]$ then $c_1 \geq c_2$.

Example.

If $a_1 < a_2 < a_3$ then $[a_1, a_2]$, $[a_1, [a_1, a_2]]$, $[a_2, [a_1, a_3]]$ are basic, but $[a_2, a_1]$, $[a_1, [a_2, a_3]]$ are not.

In the free group $F(A)$ we can interpret $[c_1, c_2]$ as $c_1 c_2 c_1^{-1} c_2^{-1}$, etc. We now answer the following:

4.3.4 How many basic commutators are there?

We consider strings and cycles, $c_1 c_2 \ldots c_n$ of basic commutators, reading the strings from left to right and the cycles clockwise as in Figure 3.

So in particular, $c_1 c_2 \ldots c_n$, $c_2 c_3 \ldots c_n c_1$, $c_3 \ldots c_n c_1 c_2$ etc. are the same cycle.

We say that a string or cycle $c_1 c_2 \ldots c_n$ is **decreasing** if $c_1 \geq c_2 \geq \ldots \geq c_n$. An immediate consequence of the definition is

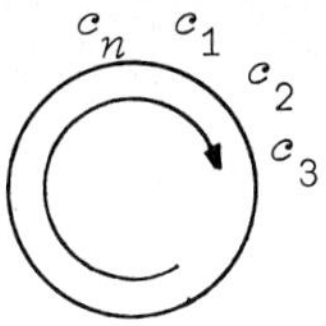

Figure 3. The cycle of commutators $c_1 c_2 \dots c_n$.

Lemma 4.3.5

A decreasing cycle has all its members equal.

Write $c_k \lhd c_{k+1}$ if $[c_k, c_{k+1}]$ is basic. A **bracketing** of a string or cycle is a sequence of moves of the form

$$c_1 c_2 \dots c_k c_{k+1} \dots c_n \longrightarrow c_1 c_2 \dots [c_k, c_{k+1}] \dots c_n$$

where $c_k \lhd c_{k+1}$.

The bracketing is called **decreasing** if the resulting string or cycle is decreasing.

Lemma 4.3.6

If a string starts $c_1 \lhd c_2$ *then the basic commutator* $[c_1, c_2]$ *appears in any decreasing bracketing of* $c_1 c_2 \dots c_n$. *Similarly, if* $c_{k-1} \geq c_k \lhd c_{k+1}$ *are three consecutive terms, then* $[c_k, c_{k+1}]$ *appears in any decreasing bracketing.*

Proof.

Suppose that $c_1 \lhd c_2$. Any bracketing of the string which doesn't include $[c_1, c_2]$ must be bracketed $c_1[c_2, c']$ for some basic commutator c'. But this bracketing cannot be decreasing.

Now suppose that $c_{k-1} \geq c_k \lhd c_{k+1}$. Any bracketing in which

$[c_k, c_{k+1}]$ does not appear must contain one of

(i) $[c'_{k-1}, c_k]c_{k+1}$, $c'_{k-1} \geq c_{k-1}$;

(ii) $c_{k-1}[c_k, [c_{k+1}, c'_{k+2}]]$;

(iii) $c'_{k-1}\, c_k\, c'_{k+1}$, $c'_{k-1} \geq c_{k-1}$, $c'_{k+1} \geq c_{k+1}$.

Situations (i) and (ii) are illegal since the commutators are not basic, and (iii) is not decreasing.

Definition 4.3.7 **Proper strings and cycles.**

A string or cycle is **proper** if $c_k < c_{k+1} \Longrightarrow c_k \lhd c_{k+1}$. Note that all strings or cycles in A are proper.

Theorem 4.3.8 *Any proper string (P. Hall) or proper cycle (Meier-Wunderli) can be decreasingly bracketed in a unique manner.*

Proof. We proceed by induction on the length n of the string or cycle, the case $n=1$ being trivial. If $c_1 \ldots c_n$ is proper but not decreasing then $c_{k-1} \geq c_k \lhd c_{k+1}$ (or $c_1 \lhd c_2$) occurs. Then $c_1 \ldots c_{k-1}[c_k, c_{k+1}] \ldots c_n$ is also proper and has length $n-1$. Consequently there is a decreasing bracketing. Also, by 4.3.6 $[c_k, c_{k+1}]$ must occur in any bracketing and so the bracketing is unique.

Example.

Consider the string $a_2 a_3 a_2 a_1 a_1 a_3 a_1 a_2 a_1$. This has the decreasing bracketing $[[a_2, a_3], [a_2, [a_1, [a_1, a_3]]][a_1, a_2]a_1$.

Definition 4.3.9 **The Möbius function** μ .

The arithmetic function μ is defined on positive integers by the rules:

$\mu(1) = 1$, $\mu(p_1 p_2 \ldots p_s) = (-1)^s$ for distinct primes $p_1, \ldots, p_s$ and

$\mu(d) = 0$ otherwise.

Theorem 4.3.10 (Witt)

If $\#(A) = m$ *there are* $\lambda_n = \frac{1}{n} \sum_{d|n} \mu(d)\, m^{n/d}$ *basic commutators of weight* n.

Proof.

Associated with any cycle of length n and period $d\ (d \mid n)$ are exactly d strings as follows:

$$\begin{array}{lllll}
a_1 a_2 \cdots a_d a_1 \cdots\cdots & a_d \cdots\cdots & a_1 \cdots\cdots & a_d \\
a_2 \cdots\cdots a_1 a_2 \cdots\cdots & a_1 \cdots\cdots & a_2 \cdots\cdots & a_1 \\
\vdots \qquad \vdots \ \vdots & \vdots & \vdots & \vdots \\
a_d a_1 \qquad a_{d-1} a_d \cdots & a_{d-1} \cdots & a_d \cdots\cdots & a_{d-1}
\end{array}$$

Theorem 4.3.8 says that there is a $1-1$ correspondence between basic commutators of weight n and cycles of length and period n.

Since there are m^n strings of length n in $a_1 a_2 \ldots a_m$, we have $m^n = \sum_{d|n} d\lambda_d$.

A fundamental result of number theory [5] says that we can invert the above equation to obtain the required formula for λ_n.

4.4 The Magnus embedding and the lower central series.

The following lemma is a consequence of the formulae for the augmented derivatives given by 4.2.7 and 4.2.8 .

Lemma 4.4.1

Let $f \in F^{(i)}$, $g \in F^{(j)}$ *and let* I *be a string in* A.

(i) *If* $\ell(I) < i$ *then* $\varepsilon_I(f) = 0$.

(ii) *If* $\ell(I) \le \min(i, j)$ *then* $\varepsilon_I(fg) = \varepsilon_I(f) + \varepsilon_I(g)$.

(iii) *If* $\ell(I) = i + j$ *and* $I = I_1 I_2 = I_2' I_1'$ *where*

$\ell(I_1) = \ell(I_1') = i$ *and* $\ell(I_2) = \ell(I_2') = j$ *then*

$$\varepsilon_I[f, g] = \varepsilon_{I_1}(f)\,\varepsilon_{I_2}(g) - \varepsilon_{I_1'}(f)\,\varepsilon_{I_2'}(g) .$$

If c is a basic commutator let I_c be its (unbracketed) string. By Hall's theorem, 4.3.8 $I_c = I_{c'} \Rightarrow c = c'$.

We now order strings in A lexographically.

Lemma 4.4.2

(i) $\varepsilon_{I_c}(c) = 1$;

(ii) *if* $\ell(I) = \ell(c)$ *and* $I < I_c$ *then* $\varepsilon_I(c) = 0$.

Proof.

We proceed by induction on $\ell(I)$ the case $\ell(I) = 1$ being trivial.

If $\ell(I) > 1$ and $c = [c_1, c_2]$ then $I_{c_1} < I_{c_2}$. If $I = I_1 I_2 = I_2' I_1'$ as in 4.4.1, then

$$\varepsilon_I(c) = \varepsilon_{I_1}(c_1)\,\varepsilon_{I_2}(c_2) - \varepsilon_{I_1'}(c_1)\,\varepsilon_{I_2'}(c_2) .$$

If $I \le I_c$ then $I_2' < I_2' I_1' = I \le I_c = I_{c_1} I_{c_2} < I_{c_2}$ by the properties of lexographic ordering. Hence

$$\varepsilon_I(c) = \varepsilon_{I_1}(c_1)\,\varepsilon_{I_2}(c_2)$$

by induction. If $I = I_c$ then $\varepsilon_I(c) = 1.1 = 1$ by induction. If $I < I_c$ either $I_1 < I_{c_1}$ and so $\varepsilon_{I_1}(c_1) = 0$ or

$$I_1 = I_{c_1} \text{ and } I_2 < I_{c_2}$$

which implies that $\varepsilon_{I_2}(c_2) = 0$.

Corollary 4.4.3

Let $c_1, \ldots, c_{\lambda_n}$ *be the basic commutators of weight* n.

If $f = c_1^{\alpha_1} c_2^{\alpha_2} \ldots c_{\lambda_n}^{\alpha_{\lambda_n}}$ *lies in* $F^{(n+1)}$ *then* $f = 1$.

Proof.

Assume by contradiction that at least one $\alpha_i \neq 0$. Let c_i be the commutator with smallest string I_{c_i} amongst those with $\alpha \neq 0$.

Then since $f \in F^{(n+1)}$, $\varepsilon_I(f) = 0$ where $I = I_{c_i}$. On the other hand, by 4.4.2 and 4.4.1

$$\varepsilon_I(f) = \alpha_i .$$

Theorem 4.4.4 *The basic commutators $c_1, \ldots, c_{\lambda_n}$ of weight n form a basis of the Abelian group $F^{(n)}/F^{(n+1)}$.*

Proof.

We have already seen in 4.4.3 that the image of $c_1, \ldots, c_{\lambda_n}$ in the quotient form a linearly independent set. It therefore only remains to show that they also span the quotient. To that end, consider the following commutator identities.

1. $[f_1, f_2] = [f_2, f_1]^{-1}$.
2. If $f_1 \in F^{(n_1)}$ and $f_2 \in F^{(n_2)}$ then $[f_1, f_2] \in F^{(n_1+n_2)}$.
3. If $f_3 \in F^{(n_3)}$ then
   (i) $[f_1\ f_2, f_3] \equiv [f_1, f_3][f_2, f_3] \mod F^{(n_1+n_2+n_{3+1})}$;
   (ii) $[f_1\ f_2, f_3] \equiv [f_1, f_2][f_1, f_3] \mod F^{(n_1+n_2+n_{3+1})}$.
4. (i) $[f_1, f_3][f_1^{-1}, f_3] \equiv 1 \mod F^{(2n_1+n_{3+1})}$;
   (ii) $[f_1, f_3^{-1}][f_1, f_3] \equiv 1 \mod F^{(n_1+2n_{3+1})}$.
5. $[f_1, [f_2, f_3]][f_2, [f_3, f_1]][f_3, [f_1, f_2]] \equiv 1 \mod F^{(n_1+n_2+n_{3+1})}$.

The proofs of 1 - 5 can be found in Marshall Hall's book on groups, or can be deduced from the exercises at the end of this chapter. Note that 5. is a kind of Jacobi identity and that 4(i) and 4(ii) follow from 3(i) and 3(ii) by putting $f_2 = f_1^{-1}$ and $f_2 = f_3^{-1}$ respectively.

We therefore assume by induction on the weight n that the basic commutators span $F^{(n)}/F^{(n+1)}$. The result is clearly true if $n = 1$.

Let $[f, g]$ be a non basic commutator of weight $n + 1$. By rules 1 - 4 and the inductive hypothesis we can write $[f, g]$ as a product of commutators $[b, c]$ where b, c are basic and $b < c$. If $c = [d, e]$ and $[b, c]$ is not basic we must have $b < d < e$. By rule 4. we may write

$$[b, [d, e]] \equiv [d, [b, e]] \quad [e, [b, d]]^{-1} .$$

If we order the commutators lexographically, then the left-hand side of the above equation is the product of two commutators of weight $n+1$ and which are later than $[b, c]$ in the ordering. Since the number of basic commutators is finite, we can use downwards induction to obtain the required factorisation.

Corollary 4.4.5 *The element $f \in F^{(n)}$ if and only if $\varepsilon_I(f) = 0$ for all I satisfying $0 < \ell(I) < n$; (i.e. $D_n(F) = F^{(n)}$).*

Corollary 4.4.6 $\bigcap_{n=1}^{\infty} F^{(n)} = \{1\}$.

4.5 Eilenberg - Maclane spaces.

A connected space X is a $K(\pi, n)$ if

$$\pi_i(X) = \begin{cases} \pi & i = n \\ 0 & i \neq n . \end{cases}$$

Spaces such as X are called **Eilemberg-Maclane spaces.** The fundamental theorem about $K(\pi, n)$s is:

Theorem 4.5.1 *Given any integer $n \geq 1$ and group π (Abelian if $n > 1$) there is a simplicial complex K which is a $K(\pi, n)$. Moreover, for fixed π and n' any two $K(\pi, n)$'s are homotopy equivalent.*

Proof. Full details of the proof may be found in the original papers of Eilenberg and Maclane sited in the index. However, the gist of the proof runs as follows:

Firstly, construct a space Y by attaching $(n+1)$-cells to a wedge of n-spheres so that

$$\pi_i(Y) = \begin{cases} \pi & i = n \\ 0 & i < n. \end{cases}$$

Now inductively kill off higher dimensional elements of π_k by adding appropriate cells.

To see uniqueness note that the obstructions to constructing a homotopy equivalence vanish.

In what follows we shall only be interested in the case $n = 1$.

Example 4.5.2

Let Σ be a closed compact surface, $\Sigma \neq p^2$ or s^2, then Σ is a $K(\pi, 1)$ where $\pi = \pi_1(\Sigma)$.

Example 4.5.3

Let M be a 3-manifold satisfying:

1. M is closed, compact and connected;
2. $\pi_1(M)$ is infinite;
3. $\pi_2(M) = 0$.

Then M is a $K(\pi, 1)$. To see this, let $\tilde{M}$ be the universal cover of M. Then $\pi_1(\tilde{M}) = 0$ by definition and $\pi_2(\tilde{M}) = \pi_2(M) = 0$. Now $H_3(\tilde{M}) = 0$ since $\tilde{M}$ is connected and infinite, and $H_k(\tilde{M}) = 0$ $k > \dim \tilde{M} = 3$.

So by the Hurewicz theorem $\pi_i(M) = \pi_i(\tilde{M}) = 0 \quad i > 1$ and M is a $K(\pi, 1)$.

The example above illustrates the general principal that X is a $K(\pi, 1)$ if and only if its universal cover is contractible.

4.5.4 Construction of $K(\pi, 1)$ when π is $\mathbb{Z}^n$.

Let A^n denote the free abelian group with generators $a_1, \ldots, a_n$. Then $K(A^n, 1)$ is easily identified with $S^1 \times S^1 \times \ldots \times S^1$ (n times).

The universal cover is R^n and we now describe a cell subdivision $\Xi(n)$ of R^n compatible with the translations induced by elements of A^n.

Let B_i denote the oriented unit segment Θe_i where Θ is the origin in R^n and e_i is the unit point on the x_i axis.

Let $B_{i_1 \ldots i_k}$ denote the formal alternating product

$$B_{i_1 i_2 \ldots i_k} = B_{i_1} \wedge B_{i_2} \wedge \cdots \wedge B_{i_k} .$$

So

$$B_{i_1 i_2 \ldots i_k} = \mathit{sign}\,(\sigma)\, B_{\sigma(i_1)\sigma(i_2)\ldots\,\sigma(i_k)}$$

for any permutation σ of $1, 2, \ldots, k$ and

$$B_{i_1 i_2 \ldots i_k} = 0$$

if $i_1 i_2 \ldots i_k$ has any repeated suffices.

The same symbol will also denote the oriented k-cube with edges $B_{i_1}, \ldots, B_{i_k}$ provided $i_1 \ldots i_k$ are distinct suffices.

Let $a_i : R^n \longrightarrow R^n$ be the unit translation $a_i(x) = x + e_i$ along the i-axis. The vertices of $\Xi(n)$ are all of the form f where f is a monomial in the Laurent ring $\Lambda[a_1, \ldots, a_n]$. So a typical vertex of $\Xi(n)$ will be $a_1^{\nu_1} \ldots a_n^{\nu_n}$ where $\nu_1, \ldots, \nu_n$ are integers. The k-cells, $k > 0$ of $\Xi(n)$ are the translates of $B_{i_1 \ldots i_k}$.

The boundary homomorphisms

$$\partial : C_k(\Xi(n)) \longrightarrow C_{k-1}(\Xi(n))$$

are given on generators by

$$\partial B_{i_1 \ldots i_k} = \sum_{j=1}^{k} (-1)^j \, (a_j - 1) \, B_{i_1 \ldots \hat{i}_j \ldots i_k} \qquad k > 1$$

and $\quad \partial B_i = (a_i - 1) \qquad k = 1.$

Since $H_j(\Xi(n)) = 0 \quad j > 0$ we have an exact sequence

$$0 \to C_n(\Xi(n)) \to C_{n-1}(\Xi(n)) \to \ldots \to C_1(\Xi(n)) \to C_0(\Xi(n)) \xrightarrow{\varepsilon} \Lambda \to 0.$$

As a consequence of exactness at $k = 1$ we have:

Lemma 4.5.5

Let F *be the free* Λ*-module with generators* $x_1, \ldots, x_n$ $n \geq 2$ *and let* E *be the submodule of all elements* $f_1x_1 + \ldots + f_nx_n$ *in* F *satisfying* $f_1(a_1 - 1) + \ldots + f_n(a_n - 1) = 0$. *Then* E *has a presentation with* $\binom{n}{2}$ *generators*

$$x_{ij} = (a_j - 1)\, x_i - (a_i - 1)\, x_j, \quad 1 \leq i < j \leq k$$

and $\binom{n}{3}$ *relations*

$$r_{ijk} = (a_i - 1)\, x_{jk} - (a_j - 1)\, x_{ik} + (a_k - 1)\, x_{ij} = 0.$$

(This lemma is important when considering Abelian covers and links.)

Recall the definition of doodles and cobordism classes of doodles Cob_n from Chapter 2. Identifying Borromean doodles with the boundary of cubes and using exactness of $C_k(\Xi(n))$ at $k = 2$ we get:

Lemma 4.5.6 *The* Λ*-module* Cob_n *of cobordism classes of* n *coloured doodles is generated by the set of Borromean doodles* B_{ijk} *subject to the relations*

$$(a_i - 1) B_{jk\ell} + (a_j - 1) B_{k\ell i} + (a_k - 1) B_{\ell ij} + (a_\ell - 1) B_{ijk} = 0 .$$

4.5.7 Construction of $K(\pi, 1)$ **when** π **is** $\mathbb{Z}_p$.

In constructing this Eilenberg-Maclane space which we shall call L_p it is reasonable to start with the 2-complex $K = \{ a \mid a^p \}$. The universal cover consists of a circle and p discs spanning the circle but disjoint otherwise. The circle is the lift of a together with its $(p-1)$ translates. The discs are the p lifts of the relations $a^p = 1$. So $H_2(\tilde{K}) = \pi_2(K)$ is free Abelian of rank $(p-1)$. We can kill off the generators of $H_2(K)$ by adding 3-cells with boundary two disc lifts. The result is to replace the circle with a 3-sphere. We continue in this fashion and get a 5-sphere, a 7-sphere and so on, ad infinitum.

More precisely, let S^1 be subdivided by the p 0-cells $e_0, ae_0, \dots, a^{p-1} e_0$, and the p 1-cells $e_1, ae_1, \dots, a^{p-1} e_1$ with boundary $\partial a^i e_1 = a^i \partial e_1 = a^i (a-1) e_0$.

Now proceed by induction on dimension. Write the $2k+1$ sphere as the join, $S^{2k+1} = S^1 * S^{2k-1}$. Let the cell decomposition of S^{2k-1} have the same cells as S^{2k-1} in dimension $\le 2k-1$. Let the $2k$-cells be of the form

$$a^i e_0 * S^{2k-1} = a^i e_{2k} , \quad i = 0, 1, \dots, p-1$$

and let the $(2k+1)$-cells be of the form

$$a^i e_i * S^{2k-1} = a^i e_{2k+1}, \quad i = 0, 1, \dots, p-1 .$$

If $\Delta = 1 + a + \dots + a^{p-1}$ and $\nabla = a - 1$ then with appropriate orientations we have

$$\partial e_{2k} = \Delta e_{2k-1} \quad \text{and}$$

$$\partial e_{2k+1} = \nabla e_{2k} .$$

The resulting cell complex $\tilde{L}_p$ is infinite dimensional and contractible. There is an action of $\mathbb{Z}_p$ on $\tilde{L}_p$ and the orbit space $L_p = \tilde{L}_p / \mathbb{Z}_p$ is the required $K(\pi, 1)$.

4.5.8 **Group Homology.**

Associated with any group G is a space $K(G, 1)$ which is uniquely defined up to homotopy equivalence. Hence it makes sense to define

$$H_i(G) = H_i(K(G, 1)).$$

Lemma 4.5.9

$$H_1(a) = G/G^{(2)} = G \text{ made Abelian.}$$

Proof. Let $K = \{A \mid R\}$ be a presentation of G. Then K is the 2-section of a $K(G, 1)$. Clearly $H_1(K(G, 1)) = H_1(K) = G/G^{(2)}$.

Theorem 4.5.10 (Stallings)

If π is the quotient group G/N there is an exact sequence:

$$H_2(G) \longrightarrow H_2(\pi) \longrightarrow N/[N, G] \longrightarrow H_1(G) \longrightarrow H_1(\pi) \longrightarrow 0 .$$

For a proof see Hilton and Stammbach.

Corollary 4.5.11 (Hopf)

Let $1 \longrightarrow N \longrightarrow F \longrightarrow \pi \longrightarrow 1$ be a free resolution of π. Then

$$H_2(\pi) = N \cap [F, F]/[N, F] .$$

Proof. Since $H_2(F) = 0$ this follows from 4.5.10 by putting $G = F$.

4.5.12 Example.

Let $G = SL(2, 5)$, the binary icosahedral group of order 120. If N is the centre of G then N is isomorphic to the cyclic group of order 2 and G/N is the alternating group A.

Now G is the fundamental group of the dodecahedral space which is a homology 3-sphere and is covered by the real 3-sphere. So $H_2(G) = 0$ and $H_1(G) = 0$. Therefore $H_2(A_5) = \mathbb{Z}_2$.

4.6 The Alexander Module [6].

Definition 4.6.1 Using Abelian covers and working with a commutative ring affords a great simplification. If K is a cell complex and H denotes $\pi = \pi_1(K)$ made Abelian then $\mathbb{Z}H$ is a commutative ring. Let $\tilde{K}_{ab}$ be the universal Abelian cover of K. Then $H_1(\tilde{K}_{ab})$ is a $\mathbb{Z}H$ module called the **Alexander module.**

We will adopt the following notational conventions. If H is the infinite cyclic group $\mathbb{Z}$ then a generator will be denoted by t. So $\mathbb{Z}H$ is the ring of finite Laurent polynomials with integer coefficients. A typical element of $\mathbb{Z}H$ has the form $c_{-i}t^{-i} + \ldots + c_0 + c_1 t + c_2 t^2 + \ldots + c_j t^j$, where each c is an integer.

So for some integer n this has the form $t^n f(t)$, where $f(t)$ is an ordinary polynomial in t. For Laurent polynomials we shall write $f(t) \doteq g(t)$ if for some n,

$$t^n f(t) = g(t).$$

Other notations for $\mathbb{Z}H$ are $\Lambda[t]$ or $\mathbb{Z}[t, t^{-1}]$.

If H is free with more than one generator then these generators will be denoted by $x, y, \ldots$ *etc.* Typical elements of $\mathbb{Z}H$ are finite commuting Laurent polynomials in $x, y, \ldots$. The reason for the change in notation when the number of generators is greater than one is that in this case the theory differs significantly in many places from the theory for one generator.

Suppose now that K is a complex. Then $H_1(\tilde{K}_{ab})$ is the quotient

$$Ker\{G(\tilde{K}_{ab}) \longrightarrow C_0(\tilde{K}_{ab})\}/Im\{C_2(\tilde{K}_{ab}) \longrightarrow G(\tilde{K}_{ab})\}.$$

Example 4.6.2

1. Let K be the complex $\{a, b \mid a^2 = b^3\}$. Then H is $\mathbb{Z}$. An isomorphism $\pi/[\pi, \pi] \longrightarrow \mathbb{Z}$ is given by the map $\alpha \to t^3$, $\beta \to t^2$.

In the universal cover the boundary of an appropriate 2-cell lift is

$$(1 + \alpha)\tilde{a} - \alpha^2(\beta^{-1} + \beta^{-2} + \beta^{-3})\tilde{b}.$$

Mapping down to $\tilde{K}_{ab}$ this becomes

$$(1 + t^3)\tilde{a} - t^6(t^{-2} + t^{-4} + t^{-6})\tilde{b}$$

$$= (1 - t + t^2)\{(1 + t)\tilde{a} - (1 + t + t^2)\tilde{b}\}.$$

So a typical boundary is of the form

$$f(t)(1 - t + t^2)\{(1+t)\tilde{a} - (1+t+t^2)\tilde{b}\}, \quad f(t) \in \Lambda[t].$$

On the other hand $g(t)\tilde{a} + h(t)\tilde{b}$ is a cycle if

$$g(t)(t^3 - 1) + h(t)(t^2 - 1) = 0.$$

Dividing by $(t - 1)$ we get:

$$g(t)(t^2 + t + 1) + h(t)(t + 1) = 0.$$

So $(t+1)$ divides $g(t)$ and (t^2+t+1) divides $h(t)$.

This means that all cycles are of the form

$$r(t)\{(1 + t)\tilde{a} - (1 + t + t^2)\tilde{b}\}.$$

So $H_1(\tilde{K}_{ab})$ is the free $\Lambda[t]$ module of rank 1 quotiented out by the ideal generated by $1-t+t^2$.

We write $H_1(\tilde{K}_{ab}) \cong \Lambda/(1-t+t^2)$.

2. Let $K = \{a, b \mid ab^2a^{-1}b^{-1}, b^2ab^{-1}a^{-1}\}$.

Under Abelianisation $a \to t$, $b \to 1$. So the 1-cycles are generated by $\tilde{b}$ and a calculation shows that the boundaries are generated by $(2t-1)\tilde{b}$ and $(2-t)\tilde{b}$. Note that the boundary ideal is not principal; i.e. it does not have a single generator. We write

$$H_1(\tilde{K}_{ab}) \cong \Lambda/(2t-1, 2-t).$$

As an Abelian group $H_1(\tilde{K}_{ab}) \approx \mathbb{Z}_3$. The action of t is given by $t.(0, 1, 2) = (0, 2, 1)$.

3. Let $K = \{a, b, c \mid aba^{-1}cb^{-1}c^{-1}, cbab^{-1}c^{-1}a^{-1}\}$.

Then H is the free Abelian group of rank 3 with generators $a \to x$, $b \to y$, $c \to z$. To calculate $H_1(\tilde{K}_{ab})$ we note that $f\tilde{a} + g\tilde{b} + h\tilde{c}$ lies in the cycle group Z_1 if and only if $f(x-1) + g(y-1) + h(z-1) = 0$. By 4.5.5 Z_1 has a set of generators

$$u = (z-1)\tilde{b} - (y-1)\tilde{c}$$

$$v = (x-1)\tilde{c} - (z-1)\tilde{a}$$

$$w = (y-1)\tilde{a} - (x-1)\tilde{b}$$

and one relation $(x-1)u + (y-1)v + (z-1)w = 0$.

On the other hand, the boundaries of the two 2-cells are given by

$$d_1 = (1-y)\tilde{a} + (x-z)\tilde{b} + (y-1)\tilde{c}$$

$$d_2 = (yz-1)\tilde{a} + z(1-x)\tilde{b} + (1-x)\tilde{c}.$$

To see how d_1 and d_2 are related to u, v and w refer to Figure 4.

$$\text{So } d_1 = -u \quad - w$$
$$d_2 = \quad -v + zw.$$

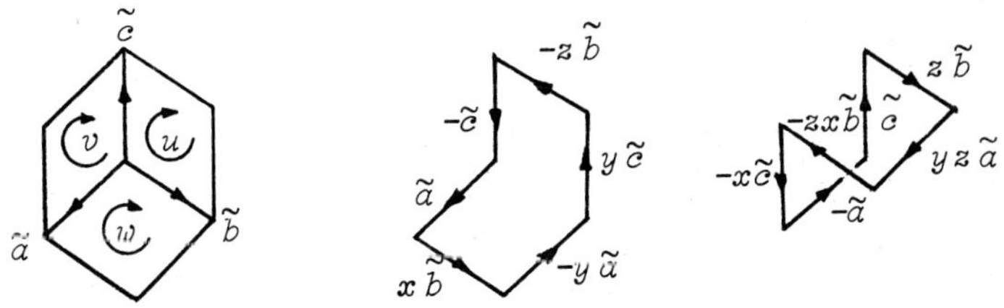

Figure 4.

We shall return to these examples after consideration of some theory.

Summing up, we have:

Theorem 4.6.3 *Let* $\tilde{K}_{ab}$ *be the universal Abelian cover of* $K = \{A \mid R\}$ *where* $H = H_1(K)$ *is free. Then* $H_1(\tilde{K}_{ab})$ *can be calculated from the chain sequence*

$$0 \longrightarrow \bigoplus_R \Lambda \xrightarrow{\partial_2} \bigoplus_A \Lambda \xrightarrow{\partial_1} \Lambda$$

where $\Lambda = \mathbb{Z}H$ *and the homomorphisms* $\partial_1, \partial_2,$ *are given by*

$$\partial_1(\tilde{a}) = 1 - \alpha \quad \text{where } \alpha \text{ is the value of } a \text{ in } H$$

and

$$\partial_2(\tilde{r}) = \sum_A (\partial_a r) \quad \text{where } \partial_a r \text{ is the value of the}$$

Fox derivative $d_a(w(r))$ *in the ring* Λ, [7].

4.7 Some Alexander module theory.

In this section, most results will be stated but not proved, [8]. A general commutative ring with identity 1 is denoted by R. The ring of Laurent polynomials in $x, y, \ldots$ will be denoted by Λ.

Definition 4.7.1

An element $u \in R$ is a **unit** if it has an inverse v (necessarily unique) such that $uv = vu = 1$.

Lemma 4.7.2

The only units in Λ *are the nomonomials* $x^n y^m \ldots$.

Definition 4.7.3

Two elements r, s in R are **associated** written $r \doteq s$ if $r = us$ for some unit u. Note that $\doteq$ is an equivalence relation and that any element f in Λ is associated to a genuine polynomial; e.g. $x^2 - xy^{-5} + 2x^{-1}y^{-1} \doteq x^3y^5 - x^4 + 2y^4$.

Definition 4.7.4

The element r **divides** the element s in R written $r \mid s$ if $s = rt$ for some t in R. If $r_1, \ldots, r_k$ are element of R then h is a **highest common factor** (h.c.f.) if $h \mid r_i$, $i = 1, \ldots, k$, and if $h' \mid r_i$ also, $i = 1, \ldots, k$ then $h \mid h'$.

Lemma 4.7.5

Any finite set of elements in Λ *has an h.c.f.*

Definition 4.7.6

An ideal $\mathfrak{J}$ in R is **principal** if $\mathfrak{J}$ is generated by a single element.

4.8 Presentation of Λ-modules and Alexander polynomials

Let $X = \{x_1, \ldots, x_n\}$ be a finite set with n elements. Write $\Lambda[X]$ for the free Λ-module of rank n generated by the elements of X.

If $A = (a_{ij})$ is an $m \times n$ matrix with entries in Λ then A is called a **Λ-matrix** and defines a presentation of the Λ-module

$$Q = \Lambda[X] / (a_{i_1} x_1 + \ldots + a_{i_n} x_n, \quad i = 1, \ldots, m).$$

In other words, Q is the Λ-module generated by $x_1, \ldots, x_n$ and with relations $a_{i_1} x_1 + \ldots + a_{i_n} x_n = 0$, $i = 1, \ldots, m$.

So returning to the examples in 4.6.2, we see that presentation matrices are

$$(1 - t + t^2), \quad \begin{pmatrix} 2t - 1 \\ 2 - t \end{pmatrix} \quad \text{and} \quad \begin{pmatrix} -1 & 0 & -1 \\ 0 & -1 & z \\ x-1 & y-1 & z-1 \end{pmatrix}$$

respectively.

In the same way that Tietze moves characterise group presentations so we have an analogous theorem for presentations of Λ-modules.

Theorem 4.8.1 *Two Λ-matrices A and B present the same Λ-module if and only if A can be changed into B by the following series of moves:*

1. *permute rows or columns;*
2. *multiply any row or column by a unit;*
3. *adjoin any row or column which is a linear combination of the original rows or columns;*
4. *change A into (from) $\begin{pmatrix} A & 0 \\ * & 1 \end{pmatrix}$ where $*$ is an arbitrary row.*

Proof. See [8].

Definition 4.8.2

In view of the above result, we search for properties of Λ - matrices invariant under these moves. The most obvious candidates involve determinants. If A is a Λ - matrix with m rows and n columns, $n \leq m$, then the ideal ε_0 generated by all $n \times n$ minors of A is called the **order ideal.** If $m < n$ take $\varepsilon_0 = (0)$. In general the i th **elementary ideal** is generated by all $(n-i) \times (n-i)$ minors if $0 < n-i \leq m$. If $n-i > m$ take $\varepsilon_i = (0)$ and if $n-i \leq 0$ take $\varepsilon_i = \Lambda$.

Since any $(n-i) \times (n-i)$ minor can be written as a linear combination of $(n-i-1) \times (n-i-1)$ minors we have $\varepsilon_i \subset \varepsilon_{i+1}$.

The i th **Alexander polynomial** Δ_i is defined to be the h.c.f. of all $(n-i) \times (n-i)$ minors, except

$$\Delta_i = \begin{cases} 0 & \text{if } n-i > m \\ 1 & \text{if } n-i \leq 0 . \end{cases}$$

Once again by expanding determinants we see that $\Delta_i \mid \Delta_{i+1}$ and that $\varepsilon_i \subset (\Delta_i)$. If ε_i is a principal ideal then and only then $\varepsilon_i = (\Delta_i)$.

To see that ε_i is invariant under the moves 1 - 4 in 4.3.1 is not difficult. However, the Alexander polynomials themselves are not invariant but may be multiplied by a unit. So if A is equivalent to A' then $\Delta_i \doteq \Delta_i'$, [9].

In the examples 4.6.2 we have respectively:

$$\Delta_i = 0, \quad i < 0, \qquad \Delta_0 = 1 - t + t^2, \qquad \Delta_i = 1, \quad i \geq 1.$$

$$\Delta_i = 0, \quad i < 0, \qquad \Delta_0 = 0 \quad \text{but} \qquad \varepsilon_0 = (2t-1,\ 2-t)$$

$$\Delta_i = 0, \quad i < 0, \qquad \Delta_0 = yz - x, \qquad \varepsilon_1 = (x-1,\ y-1,\ z-1).$$

Here are some more examples:

Example 4.8.3

1. Let $K = \{a, b \mid (ab^{-1})^{-2} b (ab^{-1})^2 a = b (ab^{-1})^{-2} b (ab^{-1})^2 \}$. Here H is $\mathbb{Z}$ generated by t the image of a and b. The

matrix $\partial_a r$ has entries $\begin{pmatrix} -2t^2 + 5t - 2 \\ 2t^2 - 5t + 2 \end{pmatrix}$,

so $\Delta_0 = 0$, $\Delta_1 = 2t^2 - 5t + 2$, $\Delta_i = 1$, $i > 1$,

and $\varepsilon_i = (\Delta_i)$ for all i.

2. Let $K = \{a,b,c \mid b^{-1}aba^{-1}b = a^{-1}ca^{-1}cac^{-1}a,\ a^{-1}cac^{-1}a = = b^{-1}cbc^{-1}b\}$.

Under Abelianisation $a = b = c = t$. The matrix $\partial_a r$ is

$$\begin{pmatrix} 3t^{-1} - 3, & -t^{-1} + 2 \\ -t^{-1} + 2, & t^{-1} - 2 \\ -2t^{-1} + 1, & 0 \end{pmatrix};$$

after row and column operations this becomes

$$\begin{pmatrix} 2 - t, & 0 \\ 0, & 1 - 2t \\ 0, & 0 \end{pmatrix}.$$

So $\Delta_1 = (2 - t)(1 - 2t) = 2 - 5t + 2t^2$,

$\Delta_2 = h.c.f.\ (2 - t,\ 1 - 2t) = 1$.

So this example has the same Alexander polynomials as the preceeding example. However, $\varepsilon_2 = (2 - t,\ 1 - 2t)$ which is not principal.

3. Now consider $K = \{\, a,b,c \mid [a,[b,c]],\ [b,[c,a]],\ [c,[a,b]]\}$. As in the third example of 4.6.2, $Z_1(\tilde{K}_{ab})$ is generated by

$$u = (z - 1)\tilde{b} - (y - 1)\tilde{c},$$
$$v = (x - 1)\tilde{c} - (z - 1)\tilde{a},$$
$$w = (y - 1)\tilde{a} - (x - 1)\tilde{b}.$$

The relations are given by the matrix:

$$\begin{pmatrix} (1 - x), & 0, & 0 \\ 0, & (1 - y), & 0 \\ 0, & 0, & (1 - z) \\ (1 - x), & (1 - y), & (1 - z) \end{pmatrix}.$$

The first three rows correspond to the three relations and can be exhibited as the boundary of a cube. The fourth row is built in to u, v, w.

Hence

$$\varepsilon_0 = ((1-x)(1-y)(1-z)),$$
$$\varepsilon_1 = ((1-y)(1-z),\ (1-x)(1-z),\ (1-x)(1-y))$$
and
$$\varepsilon_2 = ((1-x),\ (1-y),\ (1-z)).$$

4.9 Reidemeister Torsion.

Let V be a vector space over the field k and let $a = (a_1, \ldots, a_n)$, $b = (b_1, \ldots, b_n)$ be two bases for V. Let c_{ij} be the $n \times n$ matrix defined by $a_i = \sum c_{ij} b_j$. The **determinantal quotient** $[a/b]$ of the two bases is the element in $k^* = k - \{0\}$ given by

$$[a/b] = det(c_{ij}).$$

The two identities

$$[a/b][c/b] = [a/b]$$
$$[a/a] = 1$$

are easily verified.

Let G be a subgroup of k^*. We say that the bases a, b of V are **G-equivalent** if $[a/b] \in G$. Because of the identities above, we see that this is an equivalence relation.

If $0 \longrightarrow U \xrightarrow{\alpha} V \xrightarrow{\beta} W \longrightarrow 0$ is an exact sequence of vector spaces over k, then bases $a = (a_1, \ldots, a_m)$ for U and $c = (c_1, \ldots, c_n)$ for W can be combined to form a product basis ac for V by the rule

$$ac = (\alpha(a_1), \ldots, \alpha(a_m),\ \overline{\beta}(c_1), \ldots, \overline{\beta}(c_n))$$

where $\overline{\beta}(c_i)$ stands for some arbitrary choice of vector in $\beta^{-1}(c_i)$, $i = 1, \ldots, n$.

Although this choice is not well defined, we do have for all choices of bases and products the formula

$$[ac/a'c'] = [a/a'][c/c'] .$$

So, in particular, if $a \sim a'$ and $c \sim c'$ are G-equivalent, then all choices of ac and $a'c'$ are G-equivalent.

Let $C : 0 \longrightarrow C_n \longrightarrow C_{n-1} \longrightarrow \cdots \longrightarrow C_1 \longrightarrow C_0 \longrightarrow 0$ be a chain complex over k. There are short exact sequences

$$0 \longrightarrow B_i \longrightarrow Z_i \longrightarrow H_i \longrightarrow 0$$

$$0 \longrightarrow Z_i \longrightarrow C_i \longrightarrow B_{i-1} \longrightarrow 0$$

associated with C in the usual way.

Suppose C_i and H_i have preferred bases c_i and h_i. Pick a basis b_i for B_i. Then this defines a basis $b_i h_i$ for Z_i and hence a basis $b_i h_i b_{i-1}$ for C_i.

Let $\tau(C) = [b_0 h_0 b_1 / c_0]/[b_1 h_1 b_0 / c_1]/\ldots/[b_n h_n b_{n-1} / c_n]$

$$= \prod_{i=0}^{n} [b_i h_i b_{i-1} / c_i]^{(-1)^i}$$

This formula is independent of the choice of b_i because if b_i' is another choice then

$$[b_i' h_i b_{i-1}' / c_i][c_i / b_i h_i b_{i-1}] = [b_i' h_i b_{i-1}' / b_i h_i b_{i-1}]$$

$$= [b_i' / b_i][b_{i-1}' / b_{i-1}]$$

and the disparities cancel in the formula for $\tau(C)$.

The element $\tau(C)$ is called the **torsion** of C. It depends on the bases for C_i and H_i. If C arises in a geometric situation, say as the chain complex associated with a cell complex, then the basis for C_i is more or less determined by the i-cells. Moreover, if C_i is acyclic $(H_i = 0,\ i > 0)$, then the bases h_i are determined. If C is not acyclic, then it may be possible to make C acyclic by a judicious choice of coefficient field.

Example 4.9.1

Consider the lens space $L(p,q)$. Choosing a minimal cell

decomposition and considering the universal cover gives a sequence

$$0 \longrightarrow C_3 \xrightarrow{\partial} C_2 \xrightarrow{\partial} C_1 \xrightarrow{\partial} C_0 \longrightarrow 0,$$

where the C_i are all free $\mathbb{Z}_p[t]/(t^p)$ modules with one generator e_i. The boundaries are given by

$$\partial e_3 = (1 - t^q)\, e_2$$

$$\partial e_2 = (1 + t + \ldots + t^{p-1})\, e_1$$

$$\partial e_1 = (1 - t)\, e_0,$$

c.f. 4.5.7.

We can make this into a acyclic chain complex over the complex field by putting $t = w = exp(2\pi i / p)$.

The torsion is then $\Delta = (1 - w^q)(1 - w)$ which is well defined up to multiplication by w^r, $(r, p) = 1$. The modulus of the torsion is $|\Delta| = 4\,|\sin(\theta/2)\sin(q\theta/2)|$ where $\theta = 2\pi r/p$.

Definition 4.9.2 Collapses, expansions and simple homotopy.

Let L be a cell complex obtained from a cell complex K by attaching two more cells, their characteristic maps being determined by a map of triples $I^n, J^{n-1}, I^{n-1} \xrightarrow{\phi} L, L, K$, $J^{n-1} = \overline{\partial I^n - I^{n-1}}$, see Figure 5.

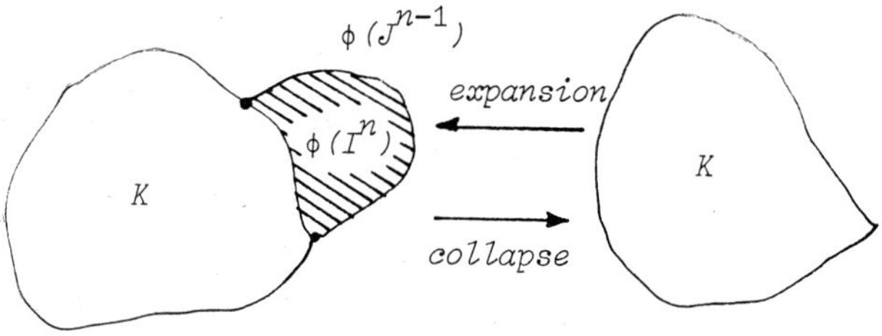

Figure 5.

The process $K \to L$ is called an **elementary expansion** of dimension n. The reverse process is called an **elementary collapse.** A sequence of expansions and collapses is called a **simple homotopy equivalence.** A simple homotopy equivalence is a homotopy equivalence but the converse is not always true. Let C, C' be two chain complexes associated with simple homotopy equivalent cell complexes. Then any preferred basis for H can be associated with a preferred basis for H'. Also, it is easily seen that the determinental quotient is unaltered by elementary expansions and collapses. For these reasons, we may state the following:

Theorem 4.9.3 *Torsion* $\tau(C)$ *is a simple homotopy invariant.*

Corollary 4.9.4 *The lens spaces* $L(7, 1)$ *and* $L(7, 2)$ *are homotopy equivalent but not simple homotopy equivalent.*

Proof. Since $2 \equiv 3^2 . 1 \mod 7$, it is well known that $L(7, 1)$ and $L(7, 2)$ are homotopy equivalent, [10]. For $L(7, 1)$ the possible values of the modulus of the torsion calculated above is, to two figures, 2.45, 3.80, 0.75. The corresponding figures for $L(7, 2)$ are quite distinct, being 1.36, 3.05, 1.69.

Comments on Chapter 4.

1. So $C_p(K^2; \mathbb{Z})$ is a π-module.

2. $\otimes_{\mathbb{Z}}$ means that the tensor product is linear over $\mathbb{Z}$.

3. An occurrence of $a_1 a_2 \dots a_k$ in the word g occurs every time that g can be written $g = \omega_1 a_1^{\eta_1} \omega_2 a_2^{\eta_2} \dots \omega_k a_k^{\eta_k} \omega_{k+1}$ $\eta_i = \pm 1$. The sign of the occurrence is $\eta_1 \eta_2 \dots \eta_k$.

4. See, for example, Marshal Hall's book.

5. See Hardy and Wright.

6. Named after the American Mathematician.

7. By the usual Roman to Greek convention.

8. See any good book on commutative rings.

9. Note that any Laurent polynomial is equivalent under $\doteq$ to a genuine polynomial.

10. See, for example, Hempel's book on 3-manifolds.

5. KNOTS AND LINKS

'There is a theory which states that if ever anyone discovers exactly what the Universe is for and why it is here it will instantly disappear and be replaced by something even more bizarrely inexplicable.
There is another theory which states that this has already happened.'

Douglas Adams — *The Hitch Hiker's Guide to the Galaxy*

This chapter is not meant, by any means, to be a comprehensive account of the theory of knots and links. For this the reader is referred to the books by Rolfsen and Crowell-Fox, or the survey article by Mc.A. Gordon. In this chapter we shall use the existence of knots and links to give beautiful examples of the mathematics developed in the previous chapters and in the next chapter we shall show how the complementary spaces of links give examples of Massey products.

Definition 5.1.1

A **link** with μ **components** is defined to be a subset ℓ of R^3 homeomorphic to the disjoint union of μ circles.

We write $\ell = \ell_1 \cup \ell_2 \cup \dots \cup \ell_\mu$ where each ℓ_i is a component of ℓ homeomorphic to a circle.

If $\mu = 1$ then ℓ is a **knot** and is usually denoted by k, [1]. It is convenient to consider R^3 and the link to be contained in $S^3 = R^3 \cup \{\infty\}$.

For each i choose an orientation of the component ℓ_i. Two links ℓ and ℓ' with the same number of components are said to be **equivalent** if there is an orientation preserving homeomorphism of spaces which restricts to an orientation preserving homeomorphism of ℓ_i into ℓ_i', $i = 1,\dots,\mu$. The notion of equivalence can be weakened somewhat if we drop the assumption of orientation preservation on either S^3 or the components ℓ_i, [2].

In order to avoid the sort of pathology illustrated in Figure 1, we shall always assume that ℓ is equivalent to a polygonal link. This is the same thing as being equivalent to a smooth C^∞ link.

Figure 1. A wild knot.

Definition 5.1.2

With this restriction every link has a **plane projection**. This is a projection of some equivalent link into a plane such that the image of the link is a collection of smooth curves which cross each other and themselves transversely. The crossing points in the plane are double points and lying above each crossing point is an **undercovering point** and above that an **overcrossing point**, see Figure 2.

The hope now is to find a representative for each equivalence class of knot or link and represent it by a plane diagram in some infinite but amenable table. Despite the discovery of some infinite families of knots and links, this programme has not been completed yet, [3].

In Figure 3, the 14 knots with less than 8 crossing points are illustrated, [4]. The first two are called the **trefoil** and the **figure eight** knot respectively.

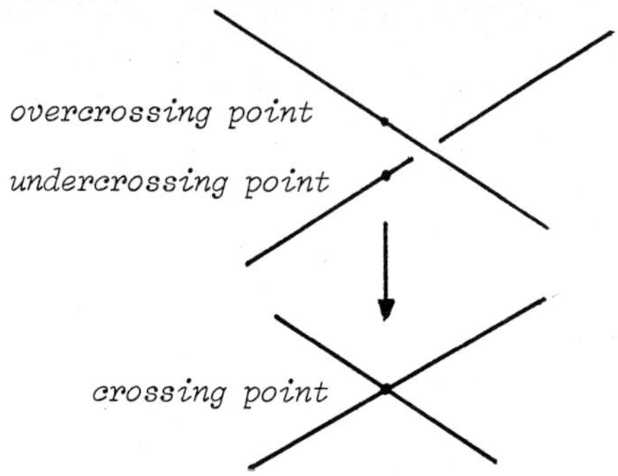

Figure 2.

Figure 3. The first 14 knots, [5].

5.2 The homology of a knot complement

As a first step towards a classification, we look at invariants of the pair (S^3, k). The most important invariant is the homeomorphism type of the space $X = S^3 - k$. The space X is called the **knot complement.** For knots it may be that the complement is a complete invariant, although this is certainly not true for links [6].

The simplest invariant of X itself is its homology groups. A simple application of Lefschetz duality implies that

$$H_i(X) \approx H_i(S^1) \approx \begin{cases} \mathbb{Z} & i = 0, 1 \\ 0 & \text{otherwise.} \end{cases}$$

So the homology groups themselves are not of much interest. In particular, they are the same for different knots. However, the manner in which H_1 is calculated is of considerable interest and gives rise to various interesting objects associated with a knot such as the spanning surface and the genus invariant.

Firstly, we consider H_1 in terms of vector fields.

5.2.1 De Rham's cohomology : mark 1

Let U be an open connected subset of S^3. Write $Rot(U)$ for the set of C^∞ vector fields A defined on U which satisfy $curl\, A = 0$ everywhere. We can extend this definition to compact subsets C of S^3 provided that we define a vector field on C to be the restriction of some vector field defined on an open neighbourhood U of C. The equation $curl\, A = 0$ is now required to be satisfied on this neighbourhood.

Elements of $Rot(U)$ are called **irrotational vector fields** and form a linear space over the real number field $\mathbb{R}$. Let $Brot(U)$ be the subspace of $Rot(U)$ consisting of all $A = grad\,\phi$ for some smooth function ϕ defined on U. Elements of $Brot(U)$ are called **exact** irrotational fields and the functions ϕ are called **scalar potential fields.**

Example 1. Let U be an open set, star convex from x_0. Then $Rot(U) = Brot(U)$. If $A \in Rot(U)$ a specific formula for a potential ϕ is given by

$$\phi(x) = \int_0^1 A(tx + (1-t)x_0) \cdot (x - x_0)\, dt .$$

This formula is an illustration of Poincaré's lemma for H^1. We shall consider the H^2 version later.

Example 2. Let $U = S^3 - \{z\text{-}axis\}$. Then $A = (x/r^2, y/r^2, 0)$ lies in $Rot(U)$ where $r^2 = x^2 + y^2$. The field A has the potential $\phi = \log r$ even though U is not star convex.

Example 3. If we rotate the previous example through $\pi/2$ we get $B = (-y/r^2, x/r^2, 0)$. This also lies in $Rot(U)$ but it does not lie in $Brot(U)$. To see this, argue as follows: Let α be the closed contour $(\cos\theta, \sin\theta, 0)$, $0 \le \theta \le 2\pi$. Integrating B around this contour gives

$$\int_\alpha B \, . \, d\alpha = \int_0^{2\pi} (-\sin\theta, \cos\theta, 0) \, . \, (-\sin\theta, \cos\theta, 0)\, d\theta$$

$$= 2\pi .$$

On the other hand, if $B = grad\,\phi$ this integral is

$$\int \frac{d\phi}{d\theta}\, d\theta = \phi(2\ ) - \phi(0) = 0 .$$

Definition 5.2.2 Periods.

Let $\alpha = [0, 1] \to U$ be a smooth curve.

The **integral** of A along α is

$$\int_\alpha A \, . \, d\alpha = \int_0^1 \left(A_1 \frac{d\alpha_1}{dt} + A_2 \frac{d\alpha_2}{dt} + A_3 \frac{d\alpha_3}{dt} \right) dt .$$

If A is irrotational the integral has been historically interpreted as the 'work' done by A in going along α.

If A is irrotational and α is a closed loop $\alpha(0) = \alpha(1)$ then $\int_\alpha A \, . \, d\alpha$ is called the **period** of A around α.

Theorem 5.2.3

An irrotational vector field is exact if and only if all its periods vanish. [7]

Proof.

Suppose $A = grad\,\phi$ then

$$\int_\alpha A\,.\,d\alpha = \int_0^1 \left(\frac{\partial\phi}{\partial\alpha_1}\frac{d\alpha_1}{dt} + \frac{\partial\phi}{\partial\alpha_2}\frac{d\alpha_2}{dt} + \frac{\partial\phi}{\partial\alpha_3}\frac{d\alpha_3}{dt} \right) dt$$

$$= \int_0^1 \frac{d\phi}{dt}(\alpha(t))\,dt \quad \text{by the chain rule}$$

$$= \phi(\alpha(1)) - \phi(\alpha(\phi)) = 0 \text{ if } \alpha(1) = \alpha(0).$$

On the other hand, suppose $\int_\alpha A\,.\,d\alpha = 0$ for all closed α. Suppose initially that U is connected and pick $u_0 \in U$. For arbitrary u in U let γ join u_0 to u. Since all periods vanish, the function

$$\phi(u) = \int_\gamma A\,.\,d\alpha \quad \text{is independent of } \gamma.$$

Moreover, differentiating gives $grad\,\phi = A$.

If U is not connected we define ϕ over each component in turn.

Lemma 5.2.4 *The period of an irrotational vector field around a loop α only depends on the homology class of α. In particular, if α is the boundary of some 2-chain in U, then $\int_\alpha A\,.\,d\alpha = 0$.*

Proof. If c is a 2-chain then Stokes' theorem [8] gives $\int_c curl\,A = \int_{\partial c} A$ from which the result follows.

Definition 5.2.5

In view of the above, we have a well-defined homomorphism Period : $H_1(U) \to \mathbb{R}$ and so every irrotational vector field defines a cohomology class in $H^1(U:\mathbb{R})$.

In order to by-pass pathology we shall now restrict our open sets U to be the interiors of compact manifolds M such that

$S^3 - U$ is a compact manifold N with interior $V = S^3 - M$ and with boundary $\partial N = \partial M = M \cap N$.

As an important example, let k be a knot and let N be a closed tubular neighbourhood of k. So N is homeomorphic to a solid torus $S^1 \times D^2$ by a homeomorphism which takes k into $S^1 \times \{0\}$.

The Mayer-Victoris sequence of the pair M, N gives

$$0 = H_2(S^3) \longrightarrow H_1(\partial N) \longrightarrow H_1(N) \oplus H_1(M) \longrightarrow H_1(S^3) = 0.$$

A basis for $H_1(\partial N)$ can be constructed as follows: Let m be a curve in ∂N which spans a disc in N. Then m is called a **meridian** of k. Let ℓ be a curve in ∂N which meets m transversely in one point. Then ℓ is called a **longitude** of k. The curves $\{\ell, m\}$ form a cycle basis for $H_1(\partial N)$. The class of m generates $H_1(M)$ and the class of ℓ generates $H_1(N)$. If k' is a closed loop in U then the homology class of k' in $H_1(U)$ is an integer multiple of $[m]$. If we use the right-hand screw rule to orient m as figure 4, then this integer $\mu(k, k')$ is called the **linking number** of k and k'.

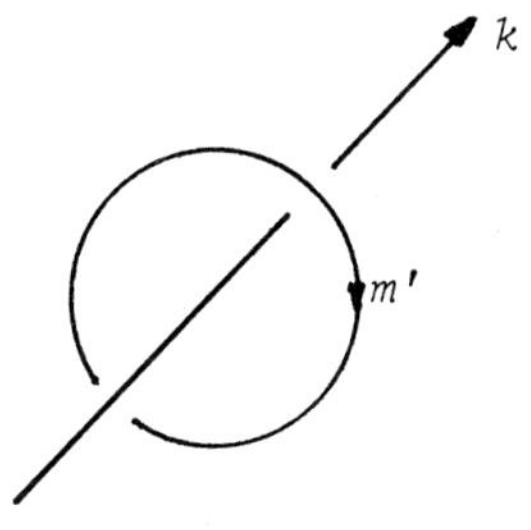

Figure 4. The right-hand screw rule.

Things to note about μ are the following:

1. μ is symmetric, so $\mu(k, k') = \mu(k', k)$.

2. For each integer n there is a k' with $\mu(k, k') = n$.

3. If c is a 2-chain with boundary k then the intersection number of k' with c is equal to $\mu(k, k')$.

Example.

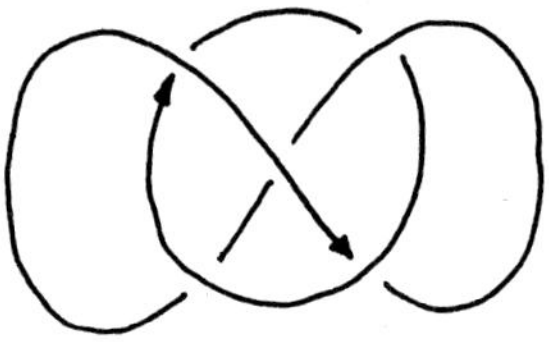

$\mu(k, k') = 2$

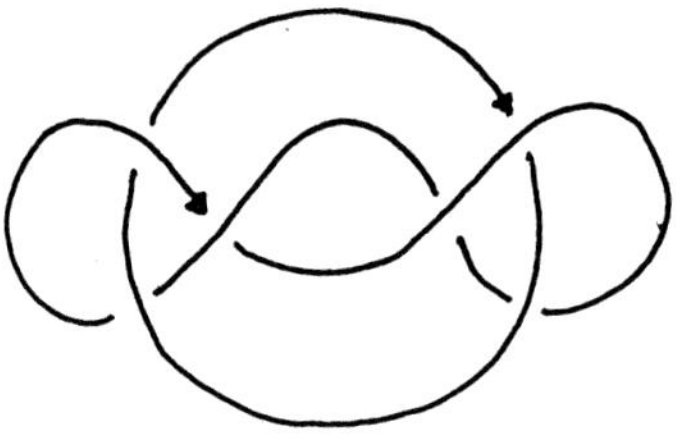

$\mu(k, k') = 0$

If k is a knot, define the vector field B_k by the formula

$$B_k(u) = \int_k \frac{dk \times (u-k)}{|u-k|^3} \quad .$$

Then by Ampère's rule [9] B_k is the electromagnetic field induced by a unit electric current flowing around k.

Since $curl\, B_k = 0$ we see that $B_k \in Rot(U)$.

Theorem 5.2.6 *The periods of* B_k *are given by*

$$\int_{k'} B_k \cdot dk' = 4\pi\mu(k, k').$$

Proof. The left-hand side is

$$\int_{k'}\int_{k} \frac{dk \times (k' - k) \,.\, dk'}{|k' - k|^3} = \int_{k'}\int_{k} \frac{(k' - k) \,.\, dk' \quad dk}{|k' - k|^3}$$

which is the change in solid angle subtended by k on traversing k', [10]. Since the solid angle changes by $\pm 4\pi$ on traversing a small meridian of k, the formula will be verified provided we check the sign. The sign is seen to be positive, on inspection of figure 5.

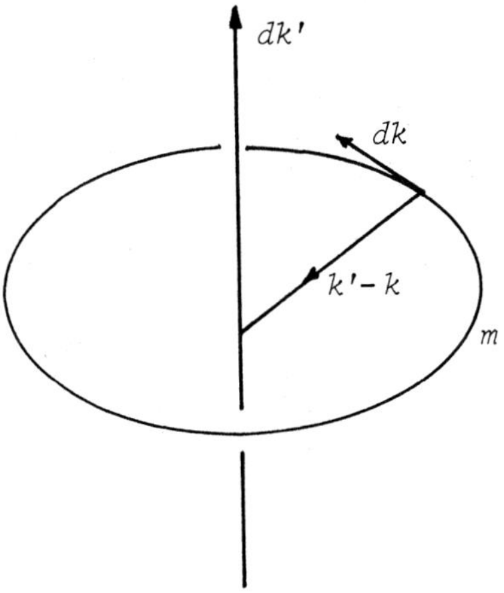

Figure 5.

Corollary 5.2.7 (Gauss).

$$\mu(k, k') = \frac{1}{4\pi} \int_{k}\int_{k'} \frac{(k - k') \,.\, dk \times dk'}{|k - k'|^3} .$$

Consider now the following situation: M, N are compact 3-manifolds with $M \cup N = S^3$ and $M \cap N = \partial M = \partial N = S$.

Theorem 5.2.8 *The groups* $H_1(N)$ *and* $H_1(M)$ *are free Abelian of the same rank. Given a cycle basis* $\{k_1, \ldots, k_n\}$ *of* $H_1(N)$ *there is a cycle basis* $\{k_1^*, \ldots, k_n^*\}$ *of* $H_1(M)$ *such that* $\mu(k_i, k_j^*) = \delta_{ij}$.

Proof.

From the Mayer-Victoris sequence of $S^3 = M \cup N$ we have

$$0 \longrightarrow H_1(S) \longrightarrow H_1(M) \oplus H_1(N) \longrightarrow 0 .$$

So $H_1(M)$ and $H_1(N)$ are free.

By Lefshetz duality, $H^1(M) = H^1(S^3 - N) \approx H_2(S^3, N) \approx H_1(N)$, and $H_1(M)$ and $H^1(M)$ have the same rank.

We can realise the isomorphism as follows: take a cycle representative k of an element in $H_1(M)$ and consider its intersection number with 2-cycles c of S^3, N. This is the same as the linking number of k with ∂c.

Corollary 5.2.9 (De Rham)

1. *If* $k_1, \ldots, k_n$ *are closed loops in* U *which form a cycle basis for* $H_1(U)$ *and if* $p_1, \ldots, p_n$ *are arbitrary real numbers, then there exists a vector field* A *in* $Rot(U)$ *such that the periods* $\int_{k_i} A$ *equal* p_i, $i = 1, \ldots, n$.

2. *Let* A_1, A_2 *be two elements of* $Rot(U)$ *with the same periods. Then* $A_1 - A_2$ *is exact. i.e.* $A_1 = A_2 + grad\,\phi$ *for some potential field* ϕ.

Proof.

1. Let $k_1^*, \ldots, k_n^*$ be dual loops as predicted by 5.2.8 and let $B_{k_i^*}$ be the electromagnetic field due to k_i^*. Then

$$A = \frac{1}{4\pi} \sum_{i=1}^{n} p_i B_{k_i^*}$$ is the required field.

2. This follows immediately from 5.2.3.

Corollary 5.2.10

$H^1(U;\mathbb{R}) \approx Rot(U)/Brot(U)$.

5.3 The Alexander module of a knot.

Let k be a knot and let N be a tubular neighbourhood of k homeomorphic to a solid torus $S^1 \times D^2$. With the usual notation, let $V = int\,N$, $M = S^3 - V$, $U = int\,M$ and $S = \partial M = \partial N$. Then $H_1(U) \approx \mathbb{Z}$, and a generator of $H^1(U;\mathbb{R})$ can be taken to be the magnetic field B_k. Locally there exists a potential field ϕ such that $grad\,\phi = B_k$. This potential is defined globally up to integer multiples of 4π. The integer multiple representing the linking number of the difference between two defining paths. By taking the values of ϕ to lie in the integers $mod\ 4\pi$ we have a well-defined function $\hat{\phi} : U \to R/4\pi\mathbb{Z} \cong S^1$.

The function $\hat{\phi}$ is real analytic and so there are only at most a discrete number of singular points for which $B = grad\,\phi = 0$. The image of the singular points in S^1 will form a finite set of singular values. Let $\theta \in S^1$ be non-singular, then $\phi^{-1}\theta \cup k$ is a smooth compact orientable surface with boundary k. To see that $\phi^{-1}\theta$ is smooth use the inverse mapping theorem and the fact that θ is a non-singular value. To see that k joins up nicely with $\phi^{-1}\theta$ to form a boundary, note that in a neighbourhood of k the level surfaces of ϕ look like the pages of a book, as in figure 6. Different pages of the book correspond to different values of ϕ.

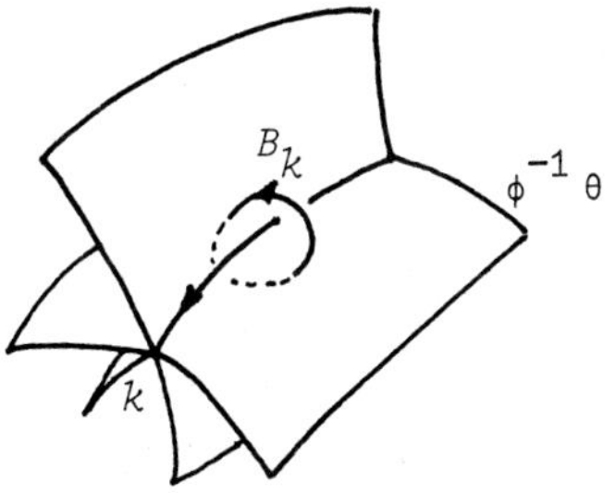

Figure 6. The form of ϕ and B_k near k.

It very often comes as a surprise when people learn for the first time that any knot can be spanned by an orientable surface. Figure 7a shows a surface (punctured torus) with boundary the trefoil knot. On the other hand, the surface in Figure 7b is non-orientable.

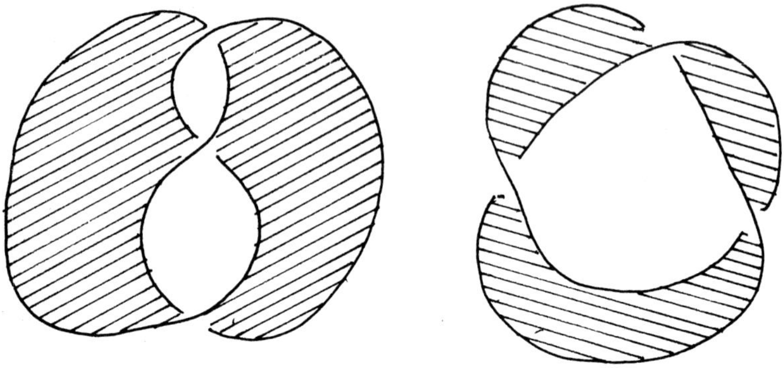

(a) (b)

Figure 7. Orientable (Seifert) and non-orientable spanning surfaces.

Definition 5.3.1 Seifert surfaces.

Any compact connected orientable surface with boundary edge the knot is called a Seifert surface [11].

Definition 5.3.2 Genus of a knot.

A Seifert surface is called minimal if its genus is minimal. The genus of a knot is defined to be the genus of a minimal Seifert surface.

Lemma 5.3.3

The genus of a knot k *is* 0 *if and only if* k *is the trivial knot.*

The genus of the trefoil knot is 1.

5.3.4 The universal Abelian cover of a knot complement.

We can now describe geometrically the Abelian cover of the

knot complement. Firstly, we need this lemma.

Lemma 5.3.5 *An element of* $\pi(S^3 - k)$ *lies in the commutator subgroup if and only if a cycle representative has zero linking number with* k.

Proof. Let α be a based loop in $S^3 - k$. Then α has zero linking number with k if and only if the homology class of α in $H_1(S^3 - k)$ is trivial. By the Hurewicz theorem this happens if and only if the homotopy class of α lies in the commutator subgroup $[\pi, \pi]$.

Corollary 5.3.6 *Let* Σ *be a Seifert surface for the knot* k. *Then a loop* α *represents an element of the commutator subgroup if and only if its intersection number with* Σ *is zero.*

Consider now the magnetic field B_k due to a knot k. The field has a locally defined potential ϕ_k defined on any simply connected subregion of $S^3 - k$, [12]. As in the theory of Riemann surfaces, we can construct a covering space $\tilde{M}$ of M on which the potential function ϕ_k is single valued. To do this consider a Seifert surface Σ equal to the component of $\phi_k^{-1}(c)$ containing k where c is a regular value of ϕ. Let C be the result of 'cutting' S^3 along Σ. Then C is a 3-manifold with boundary the union of two copies Σ_+ and Σ_- of Σ each with edge boundary k, see figure 8.

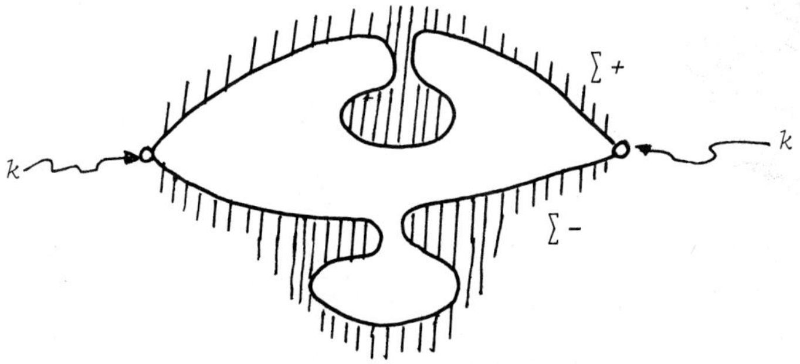

Figure 8. The cut sphere C.

We distinguish Σ_+ from Σ_- as follows. Let $x_+ \in \Sigma_+$ and $x_- \in \Sigma_-$ be formed from x in Σ. Then choose x_+ and x_- so that $\phi_k(x_+) - \phi_k(x_-) = 4\pi$.

Let $C \times \mathbb{Z}$ be a countable number of copies of C. We glue $\Sigma_+ \times \{n\}$ to $\Sigma_- \times \{n+1\}$ to form the space Y. Then $Y - k$ is a closed non-compact 3-manifold on which we can define a single-valued function $\tilde{\phi}_k : Y - k \to R$ by the rule : $\tilde{\phi}(x, n) = \phi(x) + 4\pi n$, where we have chosen a branch of ϕ single-valued on C. We therefore have a map $\hat{\phi}$ in the following commuting diagram:

$$
\begin{array}{ccc}
Y - k & \xrightarrow{\quad \tilde{\phi} \quad} & R \\
\downarrow & & \downarrow \\
S^3 - k & \xrightarrow{\quad \hat{\phi} \quad} & R\ (mod\ 4\pi R),
\end{array}
$$

where the right-hand vertical map is the universal cover of S^1 and the left-hand vertical map is the universal Abelian covering of $S^3 - k$.

Note that this construction works for any Seifert surface Σ without any need necessarily for the potential function ϕ.

We now consider a compact form of Y. Let N be a closed tubular neighbourhood of k such that $\partial N \cap \Sigma$ is a longitude ℓ. Then ∂N cut along ℓ is an annulus A. The boundary ∂A is the union of two copies ℓ_+ and ℓ_- of ℓ. The copy ℓ_+ (ℓ_-) lies in Σ_+ (Σ_-) and bounds a connected orientable surface S_+ (S_-). If $M = S^3 - N$ we can construct the universal Abelian cover $\tilde{M}$ of M in the same manner as above, see figure 9.

Let X be the 3-manifold with boundary $A \cup S_+ \cup S_-$ obtained by cutting M along S. Note that $S_\pm \cap A = \ell_\pm$. Then X lifts to a copy $X \times \{n\}$ in $\tilde{M}$. Let t denote the action of $\mathbb{Z}$ on $\tilde{M}$ so that

$$t \,.\, (X \times \{n\}) = (X \times \{n+1\}).$$

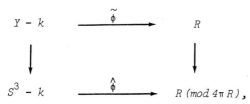

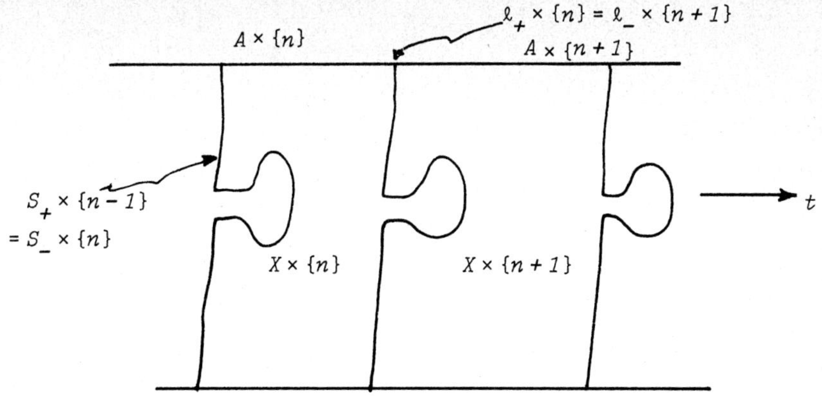

Figure 9. The Abelian cover $\widetilde{M}$.

We can now describe the Alexander module of M (and hence of $S^3 - k$) in terms of the homology of S and the action of t and hence in terms of the homology of S and the embedding of S in M.

Definition 5.3.7

Let S be an orientable surface of genus n with one boundary component. A **simplectic** spine of S is a spine consisting of the wedge of $2n$-circles $\{a_1; a_1^*, \ldots, a_n, a_n^*\}$ such that a_i, a_i^* meet transversely in the wedge point but a_i, a_j and a_i, a_j^* $i \neq j$, do not (meet transversely).

Now consider the situation above where S is a compact surface lying inside the Seifert surface Σ for the knot k.

Lemma 5.3.8 *Let* $\{a_1, a_1^*, \ldots, a_n, a_n^*\}$ *be a simplectic spine for* S, *then there exists a cycle basis* $\{b_1, b_1^*, \ldots, b_n, b_n^*\}$ *for* $H_1(X)$ *such that* $\mu(a_i, b_j) = \mu(a_i^*, b_j^*) = \delta_{ij}$ *and* $\mu(a_i, b_j^*) = \mu(a_i^*, b_j) = 0$.

Proof. By an isotopy of k we can assume that S is a disc with $2n$ ribbons attached, as in figure 10.

The cycles b_i, b_j^* are the boundaries of small discs transverse to the ribbons containing a_i, a_j^* respectively and oriented so that $\mu(a_i, b_i) = \mu(a_i^*, b_i^*) = 1$. Clearly the b_i and b_j^* have the required linking numbers. It remains to show that they form a cycle basis for $H_1(X)$.

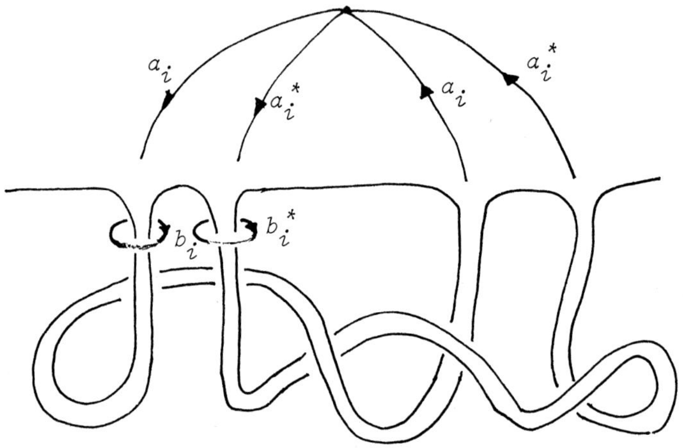

Figure 10.

By a further shrinking of S we see that X has the homotopy type of $S^3 - \{a_1, a_1^*, \ldots, a_n, a_n^*\}$. By Alexander duality $H_1(X)$ is isomorphic to $H_1(\{a_1, \ldots, a_n^*\})$ and the isomorphism is given by linking numbers.

Corollary 5.3.9 *The space X and S have the same homology.*

Definition 5.3.10 **The Seifert Pairing.**

Let $x \longrightarrow x_+$ $(x \longrightarrow x_-)$ be the map $S \longrightarrow int\, X$ which pushes points of S a little way along the positive (negative) normal vector. The **Seifert pairing** $\theta = H_1(S) \otimes H_1(S) \longrightarrow \mathbb{Z}$ is defined by the formula

$$\theta([a], [b]) = \mu(a, b_+)$$

where a, b are representative cycles in S.

Lemma 5.3.11 $\theta(a, b) - \theta(b, a) = (\theta - \theta^T)(a, b) = I(a \,.\, b)$ *the intersection number of* a *and* b *on* S.

Proof. $\theta(a, b) - \theta(b, a) = \mu(a, b_+) - \mu(b, a_+) = \mu(a, b_+) - \mu(a, b_-) = I(a \,.\, \alpha)$ where α is a ribbon with boundary $b_+ - b_-$. Finally, note that $I(a \,.\, \alpha) = I(a \,.\, b)$.

Theorem 5.3.12 *If* θ *is the Seifert pairing for a Seifert spanning surface of the knot* k *then* $\theta^T - t\theta$ *and* $\theta - t\theta^T$ *both define the Alexander invariants for* $S^3 - k$.

Proof. As we have seen earlier, the Abelian cover $\tilde{M}$ can be constructed by glueing copies of X in an infinite two-ended chain. Let $\{a_1, a_1^*, \ldots, a_n, a_n^*\}$ and $\{b_1, b_1^*, \ldots, b_n, b_n^*\}$ be the cycle bases for $H_1(S)$ and $H_1(X)$ as given by 5.3.8. Then $\{b_1, b_1^*, \ldots, b_n, b_n^*\}$ is a generating cycle set for $H_1(\tilde{M})$ as a $\Lambda[t]$ module. The relations are defined by the glueing of $S_+ \times \{i\}$ to $S_- \times \{i+1\}$ and are $(a_i)_+ = (ta_i)_-$, $i = 1, \ldots, 2n$ on the cycle level. In terms of the homology classes $[b_i]$ these become

$$\sum_{j=1}^{2n} \mu((a_i)_+, a_j)[b_j] = t \sum_{j=1}^{2n} \mu((a_i)_-, a_j)[b_j]$$

or $$\sum_{j=1}^{2n} (\theta^T - t\theta)(a_i, a_j)[b_j] = 0, \quad i = 1, \ldots, 2n.$$

A similar argument works for $\theta - t\theta^T$ by glueing $S_- \times \{i\}$ to $S_+ \times \{i-1\}$.

Corollary 5.3.13 *The Alexander invariant of a knot k has a square presentation matrix. Hence the order ideal is principal. The Alexander polynomial $\Delta(t) = \det(\theta^T - t\theta)$.*

Corollary 5.3.14 *The Alexander polynomial of a knot satisfies*

$$\Delta(t) \doteq \Delta(t^{-1}) \quad \text{and}$$
$$\Delta(1) = \pm 1.$$

Proof. $\Delta(t^{-1}) = \det(\theta^T - t^{-1}\theta) = t^{-2n}\det(t\theta^T - \theta) \doteq \Delta(t)$
$\Delta(1) = \det(\theta^T - \theta) = \pm 1$ since $\theta^T - \theta$ is the intersection form

$$\begin{pmatrix} 0 & -1 \\ +1 & 0 \end{pmatrix} \oplus \cdots \oplus \begin{pmatrix} 0 & -1 \\ +1 & 0 \end{pmatrix} \quad \text{by 5.3.11.}$$

5.3.15 **Examples.**

1. For the trefoil and the spanning surface given by figure 7 we have the Seifert form

$$\theta = \begin{pmatrix} -1 & 1 \\ 0 & -1 \end{pmatrix}.$$

From this the Alexander polynomial is given by

$$\Delta(t) = \left| \begin{pmatrix} -1 & 0 \\ 1 & -1 \end{pmatrix} - \begin{pmatrix} -t & t \\ 0 & t \end{pmatrix} \right| = t^2 - t + 1.$$

2. This example, taken from Rolfsen's book, shows that two knots may have isomorphic Alexander ideals but differing Alexander invariants.

[See Figure 11, next page.]

In (b) the Alexander module is cyclic $\Lambda[t]/(2 - 5t + 2t^2)$, but (a) has at least two generators. Note that both have $\Delta(t) = 2 - 5t + 2t^2$.

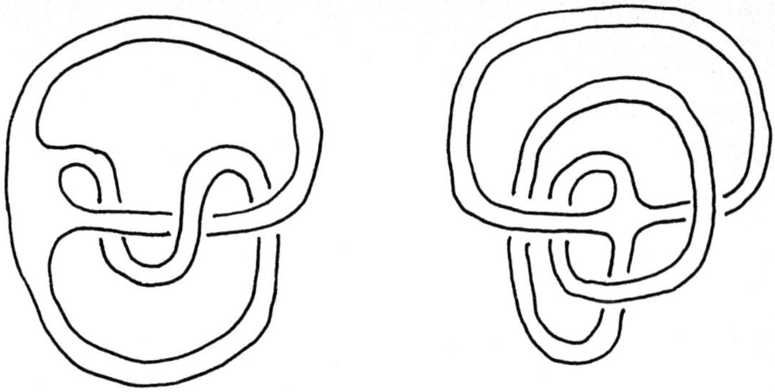

Figure 11. (a) (b)

$$\theta = \begin{pmatrix} 0 & 2 \\ 1 & 0 \end{pmatrix} \qquad \theta = \begin{pmatrix} 0 & 2 \\ 1 & 1 \end{pmatrix}$$

$$\theta^T - t\theta = \begin{pmatrix} 0, & 1-2t \\ 2-t, & 0 \end{pmatrix} \qquad \theta^T - t\theta = \begin{pmatrix} 0, & 1-2t \\ 2-t, & t-1 \end{pmatrix}$$

3. The next example illustrates a general procedure for generating knots whose Alexander module is trivial.

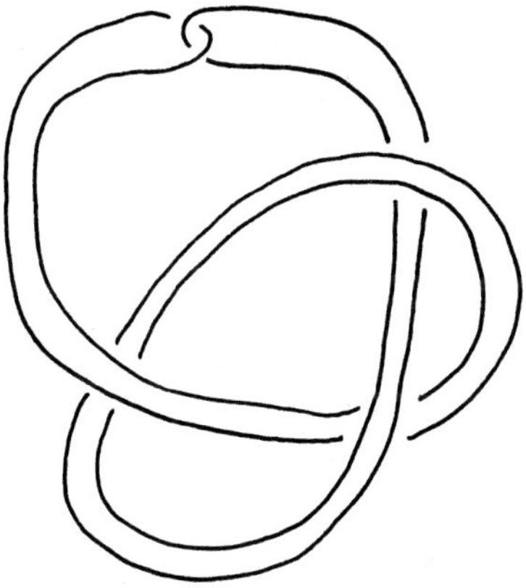

Figure 12.

Firstly, embed a ribbon in S^3 in some knotted fashion giving the ribbon enough twists so that the two boundary circles have zero linking number. Now replace a small portion of the ribbon by a clasp.

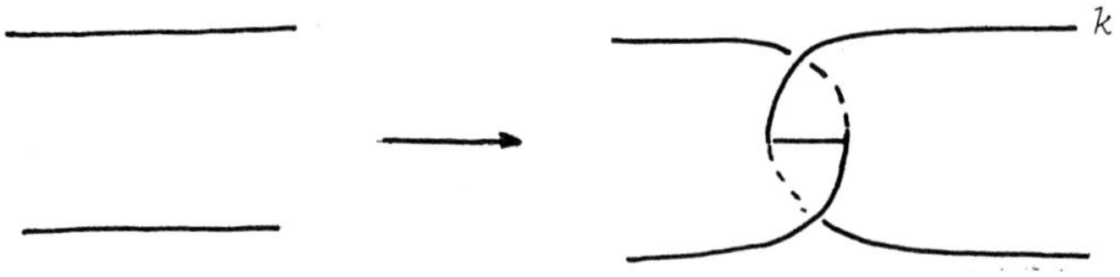

Figure 13.

This joins the boundary circles and makes a knot k. The clasped ribbon can be cut and reglued in two ways along the double arc to form a spanning surface. One regluing will give an orientable surface and the other a non-orientable surface. Choosing the orientable version defines a Seifert surface S of genus 1 with a cycle basis $a, b,$ as illustrated in figure 14.

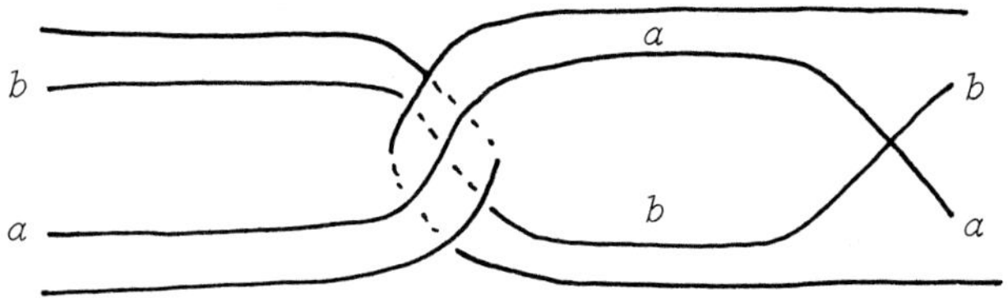

Figure 14.

Let $+$ denote a push up towards the reader. Then we see that $\mu(b, a_+) = 0 = \mu(a, a_+) = \mu(b, b_+)$ and $\mu(a, b_+) = \pm 1$.
So the Seifert matrix for this spanning surface is $\begin{pmatrix} 0 & \pm 1 \\ 0 & 0 \end{pmatrix}$ giving a trivial Alexander polynomial.

Of course we don't know yet that this method gives distinct knots. However, this fact can be deduced in many cases by considering the fundamental group of the complement, the study of which forms the content of the next section.

5.4 The fundamental group of the knot complement.

The fundamental group of $S^3 - k$ is called the **group** of k and is a very strong invariant of knots and 3-manifolds in general. Consider the following: [13].

Theorem 5.4.1 (Dehn's Lemma).

Let M be a 3-manifold such that the induced map $\pi_1(\partial M) \longrightarrow \pi_1(M)$ has non-trivial kernel. Then there exists an essential simple closed curve in ∂M which spans an embedded disc in M.

For a beautiful proof of this result see the paper by Stallings in 1958.

Corollary 5.4.2 *The knot k is trivial if and only if its group is infinite cyclic.*

Proof. If k is the flat circle $x^2 + y^2 = 1$, $z = 0$ then $S^3 - k$ is homeomorphic to $S^1 \times R^2$ where $S^1 \times (0, 0)$ corresponds to the z-axis together with ∞. So $\pi_1(S^3 - k) = \mathbb{Z}$. Clearly two equivalent knots have isomorphic groups.

To prove the converse we use Dehn's lemma above. Let k be a knot with group $\mathbb{Z}$. Let N be a tubular neighbourhood of k and let ℓ be a simple longitudinal loop in ∂N with linking number $\mu(\ell, k) = 0$. If $M = S^3 - N$ then ℓ is homotopic to zero in M since H_1 and π_1 are isomorphic in this case. So ℓ can be spanned by a singular disc in M. By adapting Dehn's lemma it is seen that ℓ can be spanned by a non-singular disc and so k is unknotted [14].

5.4.3 Finding presentations of the knot group.

It is quite easy to write down a presentation of the knot group once a projection of the knot is known. We describe the method and then give a justification later.

The n undercrossing points of a plane projection of a knot k divide k into n arcs. Let a_i label the ith arc oriented and numbered according to the orientation of k. To each double point associate one of two relations according to figure 15.

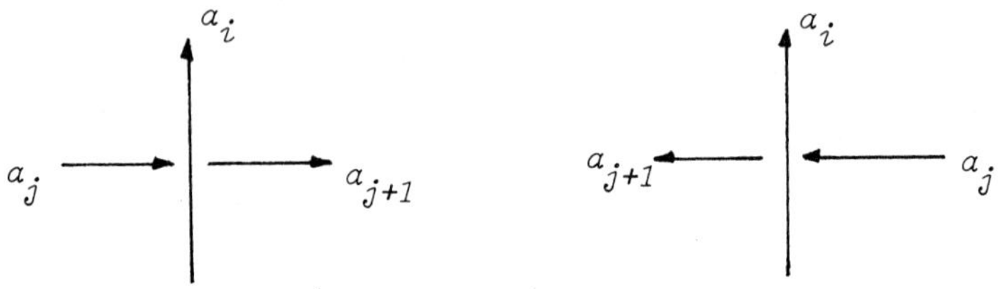

$a_{j+1} = a_i a_j a_i^{-1}$ $\qquad\qquad$ $a_{j+1} = a_i^{-1} a_j a_i$

Figure 15. The knot group relations.

Theorem 5.4.4 (The Wirtinger presentation).

With the notation above, the group of k has a presentation with n generators $a_1, \ldots, a_n$ and n relations corresponding to each double point and given by Figure 15.

Example. Consider the following projection of the figure 8 knot.

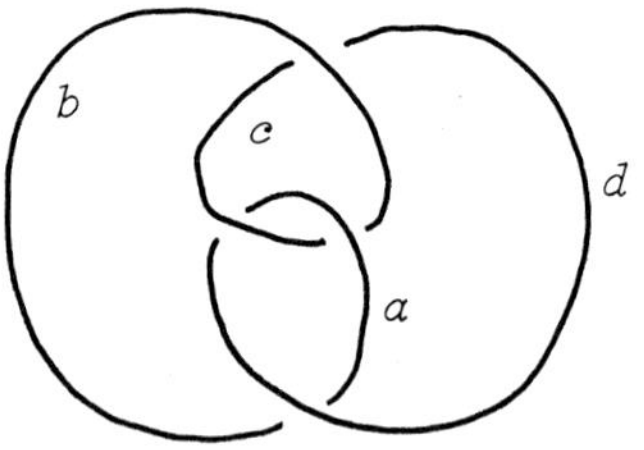

This leads to the presentation:

$$\left\{ a, b, c, d \mid a = c^{-1} d c, \quad b = d a d^{-1}, \quad c = a^{-1} b a \right\}$$

of the knot group. Note that the relation $d = b c b^{-1}$ is redundant.

5.4.5 The space M as a cell complex.

In order to prove 5.4.4 we shall now describe a cell decomposition K of the compact knot complement M. The space M is the closure of the complement of a knotted torus N. If ∞ lies in the interior of M then $M - \infty$ has as a deformation retract a 2-complex consisting of the boundary of the torus ∂N together with some two-dimensional stuff which prevents the deformation retract proceeding any further. For example, if k is the unknot then K^2 can be taken to be an unknotted hollow torus together with a disc spanning a longitude.

In general, consider a plane projection in the plane R^2. The image of the knot as a curve with self-intersections divides R^2 into a number of regions. Let E be the closure of all the bounded regions. Then E is contractible [15].

By a deformation retract we can assume that ∂N becomes a knotted tube T resting on R^2 such that $T \cap E$ is the projection of the knot and that T meets itself in meridians m_i and that the m_i meet E in the ith double point, $i = 1, \ldots, n$.

So to recap, T is a tube resting for the most part on E and meeting E in the arcs of the projection. When coming to an undercrossing T continues under an over-portion of T and when T comes to an overcrossing then T climbs over the under-portion of T meeting it in a meridian m_i before continuing onwards.

It is not hard to visualise a deformation retract of $M - \{\infty\}$ onto $K^2 = E \cup T$.

K^2 can be decomposed into a cell complex as follows. Let the vertices of K^2 be the double points of the plane projection. Add 1-cells and 2-cells to form E. Since E is contractible there is no need to keep a record of the names given to these cells. Now add 1-cells m_i to make the meridians, $i = 1, \ldots, n$. We can complete the construction of T by adding 2-cells T_i using words of the form $e_i m_{i+1}^{-1} e_i^{-1} m_i$, where e_i is some word in the other meridians, $i = 1, \ldots, n$, see figure 16.

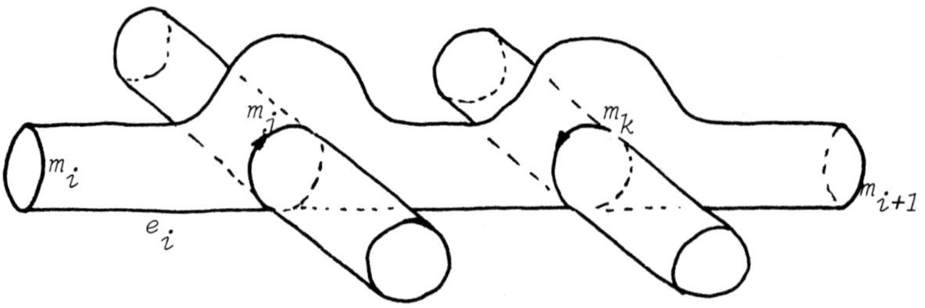

Figure 16. The 2-cell T_i. Here $e_i = m_j m_k^{-1}$.

Finally, we may add a 3-cell to K^2 to form the cell decomposition K of M. The 3-cell is attached by a map $f : S^2 \to K^2$ which covers the 2-cells of E twice and the 2-cells T_i once. To simplify matters even more, shrink the disc E to a point e. Then this doesn't change the homeomorphism type of M and shows that M has a cell decomposition with one 0-cell, n 1-cells, n 2-cells and one 3-cell.

In terms of the pictures describe in Chapter 2, the 3-cell is attached using a picture obtained from the knot projection. The 2-cells correspond to intervals between undercrossing points, as in Figure 17.

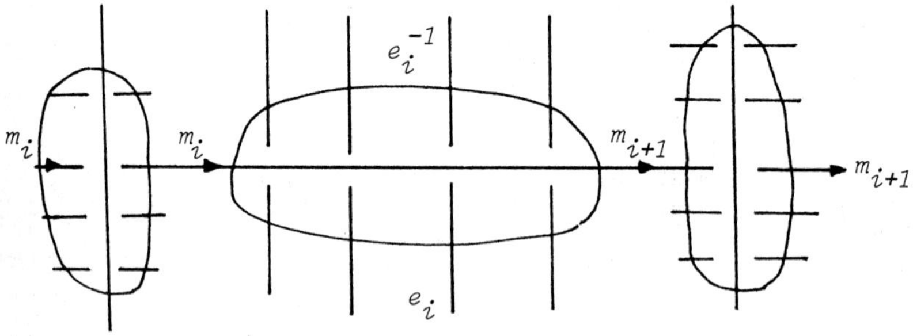

Figure 17. 2-cell discs of the 3-cell attaching map.

Proof of 5.4.4.

Now that we have described the cell complex K decomposing M we are in a position to verify that the Wirtinger presentation gives the group of the knot.

The complex immediately gives a presentation with generators $m_1, \ldots, m_n$ and relations $r_i = m_i e_i m_{i+1}^{-1} e_i^{-1} = 1$, $i = 1, \ldots, n$. If the projection is alternating, that is underpasses alternate with overpasses, then $e_i = m_j^{\pm 1}$ for some j and there is nothing more to prove. In general, we proceed as follows. Let $a_1, \ldots, a_n$ be the meridian arcs shown as in Figure 18.

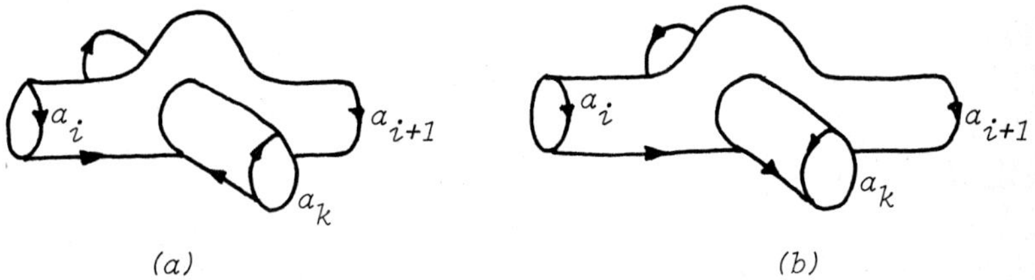

Figure 18.

Provided we agree to identify loops on both sides of the underpass (they are clearly homotopic) we get the relations

$$S_i = a_i a_k a_{i+1}^{-1} a_k^{-1} = 1 \qquad \text{(a)}$$

$$S_i = a_i a_k^{-1} a_{i+1}^{-1} a_k = 1 \qquad \text{(b)}$$

Each meridian m_j corresponds to an a_k as in Figure 18. We shall call such an a_k an **underpass loop.** Let π be the group of the knot with presentation $\{m_1, \ldots, m_n \mid r_1, \ldots, r_n\}$ and let G be the group with presentation $\{a_1, \ldots, a_n \mid s_1, \ldots, s_n\}$. We will show that π and G are isomorphic. We have already seen that corresponding to each m_j is some a_k. Let ν be the smallest positive integer such that $a_{j+\nu}$ is also an underpass loop. Then r_k is a consequence of the relations S_i , $i = j,\ldots, j+\nu$, so we have a homomorphism $\phi : \pi \to G$. Now ϕ is onto because $a_{k+\ell}$ can be written in terms of the S_i and some underpass loops $a_{k_1},\ldots, a_{k_\ell}$.

The map ϕ is also injective because any S_i depends on the r_k in which it lies. the precise relationship will depend on the expressions of a_i, a_{i+1} and a_n in terms of the m's.

We conclude by noting that G is the Wirtinger presentation, provided we view it from below.

5.4.6 Example. We illustrate the above with the following non-alternating projection.

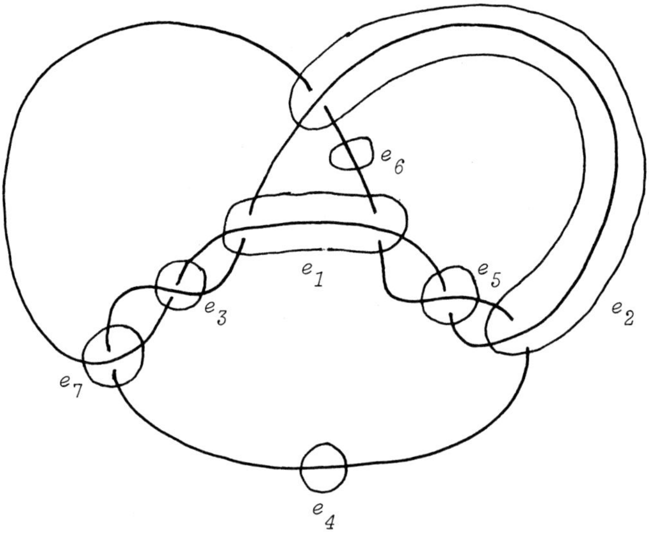

$e_1 = m_2^{-1} m_6$ $\qquad$ $m_2 = m_6^{-1} m_2 m_1 m_2^{-1} m_6$

$e_2 = m_4 m_1^{-1}$ $\qquad$ $m_3 = m_7 m_4^{-1} m_2 m_4 m_7^{-1}$

$e_3 = m_1^{-1}$ $\qquad$ $m_4 = m_1 m_3 m_1^{-1}$

$e_4 = 1$ $\qquad$ $m_5 = m_4$

$e_5 = m_2$ $\qquad$ $m_6 = m_2^{-1} m_5 m_2$

$e_6 = 1$ $\qquad$ $m_7 = m_6$

$e_7 = m_4^{-1}$ $\qquad$ $m_1 = m_4 m_7 m_4^{-1}$.

5.5 Links.

Since the theory of knots (links with one component) is only partially understood it might appear fruitless to try and understand the theory of links with more than one component. However, we can (and usually do) simplify matters by considering only those links whose components are unknotted, so that the complication lies in the interaction between different components rather than the complication inherent in each component singly. Indeed we may extend this principal so that pairwise the link is unlinked but linked by triples of components, and so on.

5.5.1 The homology of links.

The obvious first step in the understanding of links is to consider the homology of its complement. Firstly, we lay down some notation. A link with $\mu > 1$ components is denoted by $\ell = k_1 \cup k_2 \cup \dots \cup k_\mu$. A tubular neighbourhood of ℓ is denoted by N and consists of μ disjoint solid tori. The compact complement is $M = S^3 - int\, N$.

Now the first homology group of M is free on μ generators. A cycle basis could be taken by choosing a meridian from each component of ∂N.

By Alexander duality $H_2(M)$ is free of rank $\mu - 1$. Let T_i be the toroidal component of ∂N which encloses the ith component k_i of ℓ. Orient T_i so that positive normals point into k_i. Considered as a 2-cycle T_i generates a homology class λ_i in $H_2(M)$. Since T_i separates k_i from k_j $i \neq j$ and since T_i is connected, each T_i can be extended to a cycle basis for $H_2(M)$. For example $\{T_2, T_3, \dots, T_\mu\}$ is one such basis.

It is also interesting to find representative geometric cocycles for the elements of the cohomology groups of M. The first cohomology group $H^1(M)$ is isomorphic to $H_2(M, \partial M)$ and is free of rank μ. Let S_i be a compact Seifert spanning surface for k_i with boundary a longitude of T_i. Then if $j \neq i$ we can assume that each T_j meets S_i transversely in disjoint meridians of T_j spanning discs in S_i. Let S_i^0 be S_i with the interiors of these discs removed. Then S_i^0 is a relative 2-cycle in $M, \partial M$ and $\{ S_1^0, S_2^0, \ldots, S_\mu^0 \}$ is a geometric cocycle basis for $H^1(M)$. Note that this basis is dual to the meridian basis for $H_1(M)$ considered above, [16].

The second cohomology group $H^2(M)$ is isomorphic to $H_1(M, \partial M)$ and is free of rank $\mu - 1$. Let γ_{ij} be a path in M from T_i to T_j. Then $\{ \gamma_{12}, \gamma_{13}, \ldots, \gamma_{1n} \}$ is a cocycle basis for $H^2(M)$ dual to the toroidal basis for $H_2(M)$.

As in the case of H^1 we can interpret H^2 in terms of vector fields.

5.5.2 De Rham cohomology : mark 2.

Let M be a compact connected 3-manifold contained in R^3. We first note the following properties of 2-dimensional homology classes.

Lemma 5.5.3 *Any element of $H_2(M)$ can be represented by a smoothly embedded surface.*

Proof. Firstly, note that $H^1(M, \partial M) \approx H_2(M)$ is isomorphic to the set of homotopy classes $[M, \partial M; S^1, 1]$. Let $x \in H_2(M)$ correspond to a map $f : M, \partial M \to S^1, 1$. By a homotopy we can assume that f is smooth and that -1 is a regular value of f. Then $S = f^{-1}(-1)$ is the required surface.

Lemma 5.5.4 *Let M be connected with boundary components $S_0, S_1, \ldots, S_n$. Then $H_2(M)$ is free of rank M with cycle basis $\{ S_1, \ldots, S_n \}$.*

Proof. Let S be a connected closed surface in M. Then S divides M into an inside and an outside. Let $\{S_i\}_{i \in I}$ be the boundary components of M on the inside, then

$$[S] = \sum_{i \in I} [S_i].$$

In general, any homology class in $H_2(M)$ can be represented by a sum of connected surfaces and hence by a linear combination of boundary surfaces.

Since M is connected the only non-trivial relationship is

$$\sum_{j=0}^{n} [S_j] = 0.$$

Definition 5.5.5 If U is an open subset of R^3 a smooth vector field X defined on U is **solenoidal** if $\operatorname{div} X = 0$ everywhere on U. As usual, we can extend this definition to arbitrary subsets of R^3, provided X is defined on an open neighbourhood.

Let $Sol(U)$ denote the set of all solenoidal vector fields on U. Then $Sol(U)$ is a vector space over $\mathbb{R}$. The solenoidal vector field X is **exact** if $X = \operatorname{curl} A$ for some smooth vector field A defined on the **whole** of U. We call A a **vector potential** for X. The set $B\,sol(U)$ of all exact solenoidal vector fields is a subspace of $Sol(U)$.

5.5.6 **Example.** If $a \in R^3$ let X_a be the vector field

$$X_a(x) = (x - a) / \| x - a \|^3 .$$

Then X_a lies in $Sol(R^3 - \{a\})$ and is called the **flux field** out of a, [17].

Definition 5.5.7 If $X \in Sol(U)$ and S is a smooth closed surface in U then we can define the integral of X over S by the formula

$$\int_S X = \int_S (X \cdot n)\, d\Sigma$$

where n is a unit normal and $d\Sigma$ is an area infinitesimal.

Lemma 5.5.8

(i) if S and S' are homologous surfaces in U then

$$\int_S X = \int_{S'} X \,;$$

(ii) if $X \in B\,sol\,(U)$ then $\int_S X = 0$.

Proof. (i) Let $\partial V = S - S'$ then

$$\int_{S-S'} X = \int_{\partial V} X = \int_V div\, X \qquad \text{(by Gauss' theorem)}$$

$$= 0\,.$$

(ii) If $X = curl\, A$ then

$$\int_S X = \int_S curl\, A = \int_{\partial S} A \qquad \text{(by Stokes' theorem)}$$

$$= 0, \quad \text{since } \partial S = \emptyset\,.$$

Definition 5.5.9 $\int_S X$ is called the **period** of X over S.

From 5.5.8 we see that the period only depends on the homology class of S and the class of X in $Sol\,(U)\,/\,B\,sol\,(U)$.

Lemma 5.5.10 *Let X_a be the flux field out of a and let S be a smooth closed surface in $R^3 - \{a\}$. Then*

$$\int_S X_a = \begin{cases} 4\pi & \text{if } a \text{ lies inside } S \\ 0 & \text{if } a \text{ lies outside } S\,. \end{cases}$$

Proof. Let σ be the sphere $\| x - a \| = 1$ then

$$\int_\sigma X_a = \int_\sigma (x-a) \cdot (x-a)\, d\Sigma = \int_\sigma d\Sigma = 4\pi .$$

The result now follows by 5.5.8.

Lemma 5.5.11 (Poincaré : mark 2)

Let U be an open region in R^3 star convex from a. Then $Sol(U) = B\,sol(U)$; i.e. if X is a solenoidal vector field defined on U then there is a vector potential A defined on U such that $curl\ A = X$.

Moreover, if A' is any other vector potential then $A' = A + grad\ \phi$ for some scalar field ϕ.

Proof. Define A explicitly by the formula

$$A(x) = \int_{\lambda=0}^{\lambda=1} X(\lambda x + (1-\lambda)a) \times (\lambda x + (1-\lambda)a)\, d\lambda .$$

Notice that A is well defined since U is star convex from a. It can easily be verified that $curl\ A = X$.

If A' is any other solution, then

$$curl(A - A') = X - X = 0 \quad \text{so} \quad A - A' = grad\ \phi$$

by the Poincaré lemma mark 1, given in Example 1, 5.2.1.

Theorem 5.5.12 (De Rham : mark 2)

Let M be a connected compact manifold in R^3. Let $S_1, \ldots, S_n$ be n-surfaces in M which form a cycle basis for the homology group $H_2(M)$.

1. Given n real numbers $p_1, \ldots, p_n$ there is a solenoidal vector field X in M such that the period of X over S_i is p_i, $i = 1, \ldots, n$.

2. *If* Y *is a solenoidal vector field in* M *whose periods all vanish then* Y *has a vector potential* A *defined over the whole of* M.

Proof.

1. Let $\Sigma_1, \ldots, \Sigma_n$ be the inner boundary components of M. Then $\Sigma_1, \ldots, \Sigma_n$ form a cycle basis for $H_2(M)$ by 5.5.4. Let a_i be a point inside Σ_i (i.e. lying in the component of $R^3 - \Sigma_i$ which does not contain ∞), $i = 1, \ldots, n$. Then for $j \neq i$, a_i lies outside Σ_j. So the linear combination of the flux fields $X = \frac{1}{4\pi} \sum_{i=1}^{n} p_i X_{a_i}$ is an appropriate choice if $S_i = \Sigma_i$, $i = 1, \ldots, n$.

Now suppose $\left\{ S_i \right\}_{i=1}^{n}$ is an arbitrary cycle basis for H_2. Then the homology classes satisfy $[S_i] = \sum_{j=1}^{n} Q_{ij} [\Sigma_j]$ for some unimodular matrix (Q_{ij}). Let (P_{ij}) be the inverse of (a_{ij}) and let $\lambda_i = \frac{1}{4\pi} \sum_{j=1}^{n} P_{ij} p_j$, $i = 1, \ldots, n$. Then $X = \sum_{i=1}^{n} \lambda_i X_{a_i}$ is solenoidal and has the required properties.

2. The proof of this part is more complicated and will involve the use of various sublemmas. The first is of interest in its own right.

Lemma 5.5.13 *Let* K *be a cell complex and let* $\{C_p, \partial\}$ *be its associated chain complex over* $\mathbb{Z}$. *Suppose* G *is a coefficient group and* ξ *is a cochain in* $C^p(k; G)$. *Then*

(a) ξ *is a cocycle if and only if* ξ *vanishes on all boundaries in* C_p.

(b) *If* K *has free homology with* G *coefficients then* ξ *is a coboundary if and only if* ξ *vanishes on all cycles in* C_p.

Proof of 5.5.13

(a) Suppose ξ is a cocycle. Then $\xi(\partial c) = \delta\xi(c) = 0$. Conversely, suppose $\xi(\partial c) = 0$ for all $c \in C_p$. Then in particular $\xi(\partial\sigma) = \delta\xi(\sigma) = 0$ for all $(p+1)$-cells σ. So $\delta\xi = 0$.

(b) If $\xi = \delta\eta$ and z is a cycle then $\xi(z) = \delta\eta(z) = \eta(\partial z) = 0$. Conversely, and this is the hard part, let $\xi(z) = 0$ for all cycles z. Define the homomorphism $f : B_{p-1} \to G$ by $f(\partial c) = \xi(c)$. This is well defined since $\xi(c - c') = 0$ if $\partial c = \partial c'$ by hypothesis. If we can extend f to a homomorphism $F : C_{p-1} \to G$ then $F \in C^{p-1}(K; G)$, $\delta F = \xi$ and we are home.

From the short exact sequence

$$0 \longrightarrow B_{p-1} \longrightarrow Z_{p-1} \longrightarrow H_{p-1} \longrightarrow 0$$

we have the exact sequence

$$0 \longrightarrow Hom(H_{p-1}, G) \longrightarrow Hom(Z_{p-1}, G) \longrightarrow Hom(B_{p-1}, G) \longrightarrow 0,$$

since $Ext(H, G) = 0$.

So f can be extended to Z_{p-1}.

But the exact sequence

$$0 \longrightarrow Z_{p-1} \longrightarrow C_{p-1} \longrightarrow B_{p-2} \longrightarrow 0$$

splits because B_{p-2} is free. So f can be extended to C_{p-1}, as required.

Lemma 5.5.14 *Let σ be a smooth k-simplex in R^3, $k = 0, 1, 2$ or 3, and let X be a solenoidal vector field defined on σ. Suppose that there is a vector potential A for X defined on $\partial\sigma$. Then if $k \neq 2$ or if $k = 2$ and $\int_{\partial\sigma} A = \int_\sigma X$ then A can be extended to a vector potential for X over σ.*

Proof of 5.5.14

There is some vector potential A' for X since σ is diffeomorphic to a star convex region. We look at the difference between A and A' on $\partial\sigma$. Now $curl\,(A - A') = X - X = 0$ on $\partial\sigma$. If $dim\,\sigma \neq 2$, $\partial\sigma$ has $H_1 = 0$. So by De Rham mark 1 there is a scalar potential ϕ defined on $\partial\sigma$ such that $A = A' + grad\,\phi$. Extend ϕ to a differentiable function over the whole of σ. Then $A' + grad\,\phi$ is the required extension of A.

If $dim\,\sigma = 2$, then $\int_{\partial\sigma} (A - A') = \int_{\sigma} X - \int_{\partial\sigma} A' = 0$ by hypothesis and Stokes' theorem.

So by another application of De Rham's theorem $A - A' = grad\,\phi$ for some scalar field ϕ defined on $\partial\sigma$ and the same argument as in the case above applies.

Proof of 5.5.12

We now return to the proof of 5.5.12. Let Y be a solenoidal vector field with vanishing periods in M. Let K be a triangulation of M with smooth simplexes. We shall define the vector potential A step by step over the simplexes of K. By 5.5.14 there is no obstruction to this if the dimension of the simplexes is < 2. Our efforts will now be directed to overcoming this critical dimension.

Let ξ be the cochain in $C^2(K; \mathbb{R})$ defined by $\xi(\sigma) = \int_{\sigma} Y$. If z is a 2-cycle then $\xi(z) = \int_{z} Y = 0$ by hypothesis. Since homology with $\mathbb{R}$ coefficients is free we deduce that ξ is a coboundary, by 5.5.13. Let $y \in C^1(K; \mathbb{R})$ be a cochain with $\delta\eta = \xi$.

Let A' be an arbitary potential for Y defined over the 1-skeleton of K. We know by 5.5.14 that A' can be extended over the 2-section provided $\int_{\partial\sigma} A' - \int_{\sigma} Y = 0$ for each 2-simplex σ.

Now suppose that A is some potential for Y such that $\int_\tau A = \eta(\tau)$ for each 1-simplex τ. Then

$$\int_{\partial\sigma} A' - \int_\sigma Y = \eta(\partial\sigma) - \xi(\sigma) = 0,$$ so A can be extended.

We now modify A' so that it satisfies $\int_\tau A = \eta(\tau)$.

By De Rham mark 1 there is an irrotational vector field X defined over the 1-section of K such that $\int_\tau X = \eta(\tau) - \int_\tau A'$.

Let $A = A' + X$. Then $curl\ A = curl\ A' + 0 = Y$ and A satisfies $\int_\tau A = \eta(\tau)$.

So A can be extended over the 2-section and hence over the whole of K.

Corollary 5.5.15

The homology group $H^2(M; \mathbb{R})$ *and the quotient space* $Sol\,(M)/B\,sol\,(M)$ *are isomorphic* [18].

Theorem 5.5.16

If the vector fields X_1, X_2 *in* $Rot\,(U)$ *represent classes* ξ_1, ξ_2 *in* $H^1(U; \mathbb{R})$, *then* $X_1 \times X_2$ *lies in* $Sol\,(U)$ *and represents* $\xi_1 \cup \xi_2$ *in* $H^2(U: \mathbb{R})$.

Proof. By a standard formula from calculus

$$div\,(X_1 \times X_2) = (curl\ X_1) . X_2 - X_1 . (curl\ X_2).$$

If $Curl\ X_1 = curl\ X_2 = 0$ then $div\,(X_1 \times X_2) = 0$.

Now suppose we change X_1 by a gradient field $grad\,\phi$ then by another standard formula

$$(X_1 + grad\,\phi) \times X_2 = X_1 \times X_2 + curl\,(\phi X_2) - \phi\, curl\ X_2$$
$$= X_1 \times X_2 + curl\,(\phi X_2)$$

and so $X_1 \times X_2$ changes by a vector potential.

Similarly for a change in X_2. Hence $X_1 \times X_2$ represents a well-defined class in $Sol(U)/Bsol(U)$.

To see which class in H is represented by $X_1 \times X_2$ consider firstly the case where $X_i = B_i$ is the magnetic field due to disjoint closed loops k_i, $i = 1, 2$. Let T be a hollow torus surrounding k_1. We consider the integral

$$I = \int_T B_1 \times B_2 .$$

By Gauss' theorem it is irrelevant which torus we take, so we think of T as being very thin with meridian radius arbitrarily small. Let n be a unit normal to T and let $d\Sigma$ be the surface infinitesimal. Then the above integral I is

$$I = \int_T (n \times B_1) . B_2 \, d\Sigma .$$

As a surface integral we can evaluate I by firstly integrating round a meridian and then round a longitude. During the meridian motion B_2 hardly changes and the first factor integrates to 4π as we have seen in 5.2. So up to sign

$$I = 4\pi \int_{k_1} B_2 = \mu(k_1, k_2).$$

This is of course the same as the Kronecker evaluation of $\zeta_1 \cup \zeta_2$ on the torus T.

The general case follows by linearity and the fact that the magnetic fields form a cycle basis for $H^1(U:\mathbb{R})$.

5.6 The Alexander module of a link.

Under Abelianisation the group of a link with μ components becomes $\mathbb{Z}^\mu$. Let Σ_i be a Seifert surface spanning the ith component of the link ℓ. Then a loop in $S^3 - \ell$ will lift to a loop in the universal Abelian cover if and only if it has intersection number zero with each Σ_i. The action of $\mathbb{Z}^{\mu}$ on the covering space is freely generated by x_i, $i = 1,\ldots,\mu$ where x_i represents the action of crossing Σ_i transversely in a positive direction.

Although the Alexander module of a link is more complicated than that of a knot, it is still true that the order ideal is principal and that an Alexander polynomial can be defined in the variables $x_1,\ldots,x_{\mu}$.

Here are some examples:

5.6.1 Examples of Alexander modules of links.

1. Let $\ell = k_0 \cup k_1$ be two unknotted unlinked circles (the trivial link of two components). If S_1, S_2 are two disjoint spanning discs then cutting along S_1 and S_2 gives a manifold X with four open discs in its boundary. So X is $S^2 \times I$ with a circle C_0 in $S^2 \times \{0\}$ and a circle C_1 in $S^2 \times \{1\}$ missing. Let C_0 correspond to k_0 and C_1 correspond to k_1. Let $S^2 \times \{0\} - C_0 = D_0^+ \cup D_0^-$ and $S^2 \times \{1\} - C_1 = D_1^+ \cup D_1^-$ be the four discs. Take a double-countable number of copies of X say $\left\{X \times (m, n)\right\}_{m,n \in \mathbb{Z}}$. Glue $D_0^+ \times (m, n)$ to $D_0^- \times (m+1, n)$ and $D_1^+ \times (m, n)$ to $D_1^- \times (m, n+1)$. This constructs the Abelian cover. The Alexander module is a free $\Lambda(x, y)$ module of rank 1, where $x(X \times (m, n)) = X \times (m+1, n)$ and $y(X \times (m, n)) = X \times (m, n+1))$.

2. Let $\ell = k_1 \cup k_2$ be two unknotted simply linking circles, (the Hopf link).

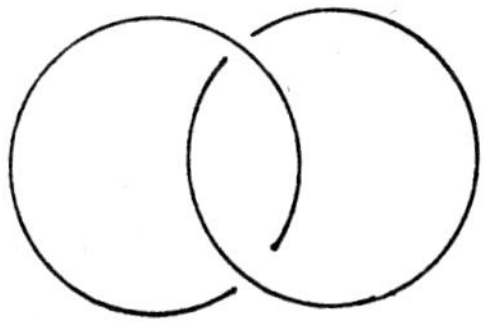

The complementary space has the homotopy type of a torus. So the Abelian cover is the same as the universal cover and is in fact homeomorphic to R^3.

Therefore the Alexander module is zero. Note that this is a less complicated example than the trivial link above!

3. Let $\ell = k_1 \cup k_2 \cup k_3$ be the Borromean rings, see Figure 19.

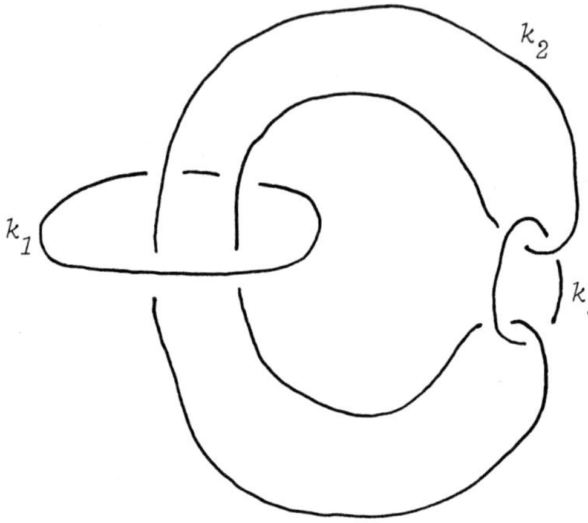

Figure 19. The Borromean rings.

Perhaps the easiest way to calculate the Alexander module is to work directly from the group presentation. Standard calculations give

$$\pi = \left\{ x, y, z \mid x, [z, y^{-1}]] , [y, [x, z^{-1}]] , [z, [y, x^{-1}]] \right\}$$

(see next section).

This gives an Alexander matrix

$$\left\{ \begin{array}{ccc} 0 , & -(1-x)(1-z)y^{-1} , & -(1-x)(1-y^{-1}) \\ -(1-y)(1-z^{-1}) , & 0 , & -(1-y)(1-x)z^{-1} \\ -(1-z)(1-y)x^{-1} , & -(1-z)(1-x^{-1}) , & 0 \end{array} \right\}$$

So $\Delta = (1-x)(1-y)(1-z)$.

5.7 Metabelian invariants of the link group.

The fundamental group of the link complement can be calculated from a plane projection in the same way as for a knot. The proof of 5.4.3 carries over in the same fashion, provided the projection is connected and this can always be arranged, [19].

As for a knot, the group of a link is a powerful invariant. The following result shows just how powerful. Call a link ℓ geometrically splittable if there is a sphere $\Sigma \subset S^3 - \ell$ such that ℓ has at least one component in each component of $S^3 - \Sigma$.

Theorem 5.7.1 *The link ℓ is geometrically splittable if and only if its group can be written as a non-trivial free product.*

Proof. One way is a simple consequence of Van Kampen's theorem. The converse follows from the sphere theorem and a result of Higman, [20] .

Combining this with 5.4.1 gives the following:

Theorem 5.7.2 *A link is equivalent to the trivial link if and only if its group is free.*

5.7.3 The Chen Groups

Although it is easy enough to write down a presentation of the group π of a link, often the group itself is fairly intractable. So we consider the lower central series $\{ \pi^{(n)} \}$ and the quotients

Definition 5.7.4

The quotients $\pi / \pi^{(n)}$ are called the Chen groups of the link, [21].

Suppose that the group of a link has a presentation of the simple form

$$\pi = \left\{ x_1, \dots, x_\mu \mid [x_1, \ell_1], \dots, [x_\mu, \ell_\mu] \right\}$$

where $\ell_i = \ell_i(x_1, \dots, x_\mu)$ are words representing longitudes of the ith component k_i , $i = 1, \ldots, \mu$.

Then the group $\pi / \pi^{(n)}$ has the presentation

$$\pi / \pi^{(n)} = \left\{ x_1, \ldots, x_\mu \mid [x_1, \ell_1], \ldots, [x_\mu, \ell_\mu], F^{(n)} \right\}$$

where $F^{(n)}$ is the nth term in the derived series of the free group $F = F(x_1, \ldots, x_\mu)$.

5.7.5 Examples.

1. The Hopf link

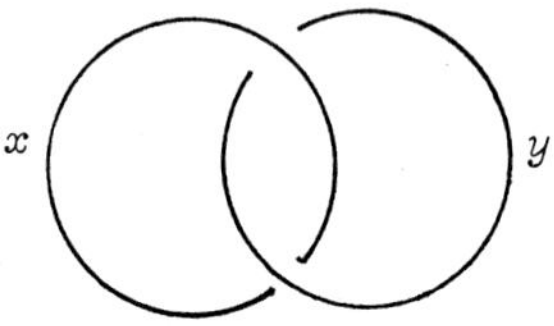

has the presentation $\{ x, y \mid [x, y] = 1 \}$, so $\pi / \pi^{(n)} \not\approx F / F^{(n)}$, $n = 2, 3, \ldots$.

2. The Borromean rings have the presentation

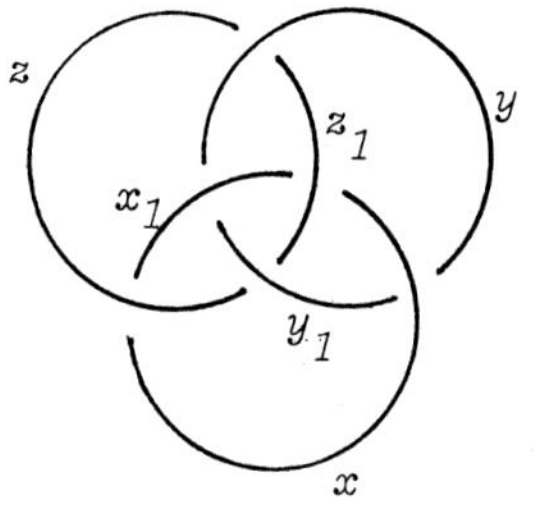

$$x_1 = z_1^{-1} x z_1 \qquad x = z x_1 z^{-1}$$

$$y_1 = x_1^{-1} y x_1 \qquad y = x y_1 x^{-1}$$

$$z_1 = y_1^{-1} z y_1 \qquad z = y z_1 y^{-1}$$

Eliminating x_1, y_1, z_1 gives the presentation:

$$\left\{ x, y, z \mid [x [z, y^{-1}]], [y, [x, z^{-1}]], [z, [y, x^{-1}]] \right\}$$

So $\pi / \pi^{(3)} \approx F / F^{(3)}$ but $\pi / \pi^{(4)} \not\approx F / F^{(4)}$ since $[x, [z, y^{-1}]] \notin F^{(4)}$. We can verify this by looking at the Magnus expansion of $[x, [z, y^{-1}]]$.

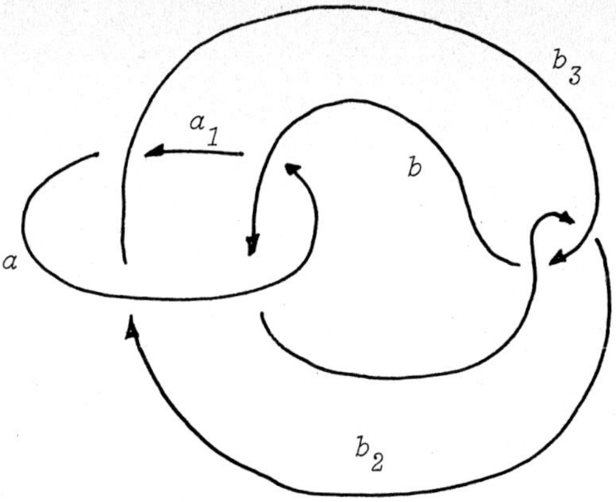

The Whitehead link has the presentation

$$a_1 = b\,a\,b^{-1} \qquad b_1 = a\,b\,a^{-1} \qquad b_3 = a^{-1}\,b_2\,a$$

$$a = b^{-1}\,a_1\,b_3 \qquad b_2 = b_3^{-1}\,b_1\,b_3 \qquad b = b_1^{-1}\,b_3\,b_1$$

$$a = b^{-1}\,b\,a\,b^{-1}\,b \qquad b = b_1^{-1}\,a^{-1}\,b_3^{-1}\,a\,b\,a^{-1}\,b_3\,a\,b_1$$

and we get the presentation

$$\pi = \left\{ a, b \mid [a, [a, b][a, b^{-1}]] \ , \ [b, [a, b^{-1}][a^{-1}, b^{-1}]] \right\}$$

Clearly $\quad \pi/\pi^{(3)} \approx F/F^{(3)}$.

Modulo $F^{(3)}$ $\quad [a, b] \equiv [a, b^{-1}]^{-1}$ and $\quad [a, b^{-1}] \equiv [a^{-1}, b^{-1}]^{-1}$

so $\quad \pi/\pi^{(4)} \approx F/F^{(4)}$.

On the other hand $\pi/\pi^{(5)} \not\approx F/F^{(5)}$ as can be seen by checking the Magnus expansion of $[a, b][a, b^{-1}] \in \pi^{(3)}$.

5.7.6 Presenting the Chen groups.

In general, the simple sort of presentation illustrated above will not be available for the group of the link. However, for presenting the Chen groups we can work under the assumption that this is so in the following sense.

Theorem 5.7.7 (Chen, Milnor).

Every link with μ *components has a sequence of presentations*

$$P^{(n)} = \left\{ a_1, \ldots, a_\mu \mid [a_1, \ell_1^{(n)}], \ldots, [a_\mu, \ell_\mu^{(n)}], F^{(n)} \right\}$$

for the Chen groups.

Proof. If a_i is the initial arc of the ith component in some plane projection let $a_{i1}, a_{i2}, \ldots, a_{i\nu_i}$ be the other arcs in turn. Then π has a presentation with generators $a_i, a_{i1}, \ldots, a_{i\nu_i}$, $i = 1, \ldots, \mu$ and relations $\{R_{ij}\}$,

$$a_{i1} = u_{i1} a_i u_{i1}^{-1}, \quad a_{i2} = u_{i2} a_{i1} u_{i2}^{-1}, \ldots, \quad a_i = u_i a_{i\nu_i} u_i^{-1},$$

where the u_{ij} are generators or inverses of generators.

The first job is to collect the generators into strings. Let $\ell_{i1} = u_i$, $\ell_{i2} = u_{i2} u_{i1}, \ldots, \ell_i = u_i u_{i\nu_i} \ldots u_{i2} u_{i1}$, then an equivalent set of relations is $\{S_{ij}\}$,

$$a_{i1} = \ell_{i1} a_i \ell_{i1}, \quad a_{i2} = \ell_{i2} a_i \ell_{i2}, \ldots, \quad a_i = \ell_i a_i \ell_i^{-1}.$$

The word ℓ_i represents a longitude along the ith component and the ℓ_{ij}, $1 \le j \le \nu_i$ represent part longitudes.

Let F be the free group on $a_1, \ldots, a_\mu$ and let Φ be the free group on a_{ij}, $1 \le i \le \mu$, $1 \le j \le \nu_i$. We will construct a sequence of homomorphisms $p^{(n)} : \Phi \longrightarrow F$ $n = 2, 3, \ldots$ and put $\ell_i^{(n)} = p^{(n)}(\ell_i)$.

The homomorphisms are constructed inductively and start with $p^{(2)}$ the constant homomorphism with image 1. Assume that $p^{(2)}, \ldots, p^{(n-1)}$ and hence the $\ell_i^{(n-1)}$ are defined. Let $\ell_{ij}^{(n-1)} = p^{(n-1)}(\ell_{ij})$ and define $p^{(n)}$ on generators by $p^{(n)}(a_{ij}) = \ell_{ij}^{(n-1)} a_i (\ell_{ij}^{(n-1)})^{-1}$. (For example, $p^{(3)}(a_{ij}) = a_i$ for all i, j).

We will now show that $p^{(n)}$ induces an isomorphism between

$\pi/\pi^{(n-1)}$ and the groups $G(n-1)$ whose presentation $P^{(n)}$ is given in the statement of this theorem. Now $\pi/\pi^{(2)}$ is free Abelian on $a_1, \ldots, a_\mu$. So $p^{(3)}$ induces an isomorphism $\pi/\pi^{(2)} \longrightarrow G(2)$.

In order to show that $p^{(n)}$ induces a homomorphism, we need to check that any relator in the presentation of $\pi/\pi^{(n-1)}$ is mapped to a consequence of the relators in $P^{(n-1)}$.

Now clearly $p^{(n)}$ takes $\Phi^{(q-1)}$ into $F^{(q-1)}$ so it only remains to check the relations of the form $a_{ij} = \ell_{ij} a_i \ell_{ij}^{-1}$. Now by definition $p^{(n)}(a_{ij}) = p^{(n-1)}(\ell_{ij}) a_i p^{(n-1)}(\ell_{ij}^{-1}) =$
$= p^{(n-1)}(\ell_{ij} a_i \ell_{ij}^{-1}) \equiv p^{(n-1)}(a_{ij})$ by induction on n say.
Hence $p^{(n)}(\ell_{ij}) \equiv p^{(n-1)}(\ell_{ij})$ and so $p^{(n)}(a_{ij}) = p^{(n)}(\ell_{ij} a_i \ell_{ij}^{-1})$.
Therefore $p^{(n)}$ induces a homomorphism which we must check is an isomorphism. The homomorphism is clearly onto because $p^{(n)}(a_{ij}) = a_i$ so it only remains to check that the kernel is a consequence of the relations in $\pi/\pi^{(n)}$. But $\left[a_i, \ell_i^{(n)}\right] =$
$= a_i p^{(n)}(\ell_i) a_i^{-1} p^{(n)}(\ell_i^{-1}) \equiv 1$ etc.

To see how this theorem works in practise, consider the following examples.

5.7.8 Examples.

1.

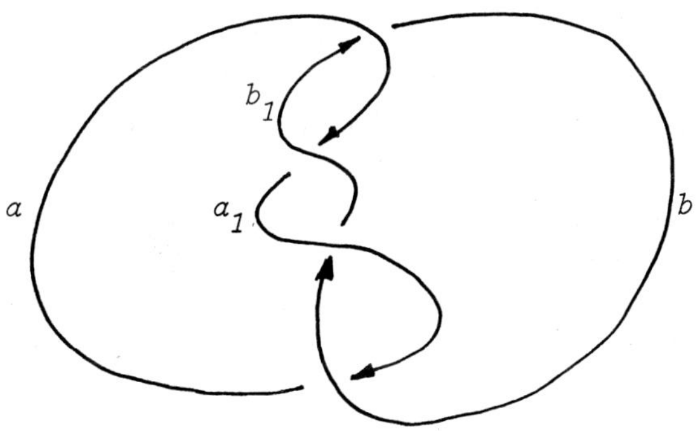

This link has presentation

$$a_1 = b_1 a b_1^{-1} \qquad b_1 = a_1 b a_1^{-1}$$

$$a = b a_1 b^{-1} = b b_1 a b_1^{-1} b^{-1} \qquad b = a b_1 a^{-1} = a a_1 b a_1^{-1} a^{-1}.$$

Then $p_3(b_1) = b \quad p_3(a_1) = a$ giving the presentation

$$\left\{ a, b \mid F^{(2)} \right\} \quad of \quad \pi/\pi^{(2)} \approx \mathbb{Z} \oplus \mathbb{Z}.$$

For $n = 4$ we get $p_4(a_1) = b^2 a b^{-2} \quad p_4(b_1) = a^2 b a^{-2}$,

so $\pi/\pi^{(3)}$ has the presentation

$$\left\{ a, b \mid [a, b^2], [b, a^2], F^{(3)} \right\}.$$

Similarly $\pi/\pi^{(4)}$ has the presentation

$$\left\{ a, b \mid [a, ba^2 b], [b, ab^2 a], F^{(4)} \right\}$$

and so on.

2. Milnor's Link.

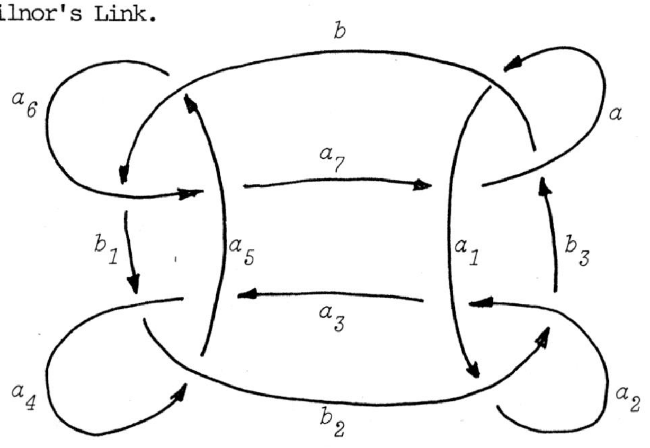

Successive right multiples of the part longitudes can be read off round a and are $b^{-1}, b_2, a_1, a_5^{-1}, b_2^{-1}, b, a_5$ and a_1^{-1}.

For b these are $a_6, a_4^{-1}, a_2, a^{-1}$.

Making the substitution $b_i \to b$ and $a_i \to a$ sends ℓ_a and ℓ_b to 1 so $\pi/\pi^{(i)} = F/F^{(i)}$ for all $i \geq 2$.

Consequently more delicate invariants are needed to prove this link non-trivial.

5.8 Generalised linking numbers : The Milnor $\bar{\mu}$ - invariants.

While calculating some of the Chen groups in the previous section we saw that $\pi/\pi^{(i)} \neq F/F^{(i)}$ by checking the Magnus expansion of a longitude. We now make this procedure systematic.

Given a presentation $P^{(n)} = \left\{ a_i \mid [\, a_i , \ell_i^{(n)}], \ F^{(n)} \right\}$ for $\pi/\pi^{(n)}$ we can assume by multiplying $\ell_i^{(n)}$ by a_i^k for some k that $\ell_i^{(n)}$ is a longitude having zero linking number with the component k_i.. A Magnus expansion $a_i \longrightarrow 1 + t_i$ expresses $\ell_i^{(n)}$ as an infinite series

$$\ell_i^{(n)} = 1 + \sum \mu\,(i_1, i_2, \ldots, i_s, i)\, t_{i_1} t_{i_2} \ldots t_{i_s} .$$

Thus an integer $\mu\,(i_1, \ldots, i_s, i)$ is defined for each sequence $i_1, \ldots, i_s, i$ of integers between 1 and μ. These integers are independent of n if $n > s$.

Let $\Delta\,(i_1, \ldots, i_r)$ denote the greatest common divisor of $\mu\,(j_1, \ldots, j_s)$ where $j_1, \ldots, j_s$ $(2 \leq s < r)$ is to range over all sequences obtained by deleting at least one of the indices $i\ , \ldots, i_r$ and permuting the remaining indices cyclically.
(For example, $\Delta\,(i j) = 0$ and

$$\Delta\,(i j k) = g.c.d. \left\{ \mu(jk),\ \mu(kj),\ \mu(ij),\ \mu(ji),\ \mu(ik),\ \mu(ki) \right\} ,$$

in fact $\mu(i j) = \mu(j i)$).

Let $\bar{\mu}(i_1 i_2 \ldots i_r)$ denote the residue class of $\mu(i_1 i_2 \ldots i_r)$ modulo $\Delta(i_1 i_2 \ldots i_r)$.

Theorem 5.8.1 *The residue classes* $\bar{\mu}(i_1 \ldots i_r)$ *are link invariants if* $r \leq n$.

Proof. Let $P^{(n)} = \left\{ a_i \mid [a_i, \ell_i^{(n)}], F^{(n)} \right\}$ be the presentation of $\pi/\pi^{(n)}$ predicted by 5.7.7.

Then the pairs $(a_i, \ell_j^{(n)})$ are well defined up to conjugation. So in order to prove 5.8.1 we need to check the following facts:

The residue class $\bar{\mu}(i_1, \ldots, i_s, i)$ is unaltered if

1. $\ell_i^{(n)}$ $(s < n)$ is replaced by a conjugate.
2. a_j is replaced by a conjugate, $j = 1, \ldots, \mu$.
3. $\ell_i^{(n)}$ is multiplied by a consequence of the words $[a_j, \ell_j^{(n)}]$, $j = 1, \ldots, \mu$.
4. $\ell_i^{(n)}$ is multiplied by an element of $F^{(n)}$.

In the Magnus power series let D_i denote the set of all elements $\sum \nu(i_1 \ldots i_s) t_{i_1} \ldots t_{i_s}$ with coefficients satisfying

$$\nu(i_1 \ldots i_s) \equiv 0 \mod \Delta(i_1 \ldots i_s\, i)$$

for all $i_1 \ldots i_s$ with $s < n$. (There is no restriction if $s \geq n$.)

Then D_i is a 2-sided ideal. For if $\nu(i_1 \ldots i_s) t_{i_1} \ldots t_{i_s}$ is a monomial in D_i and $\lambda t_{j_1} \ldots {}_{j_k}$ is arbitrary then either $s \geq n$ so that the product is trivially in D_i or $\nu(i_1 \ldots i_s) \equiv 0$ $\mathrm{mod}\ \Delta(i_1 \ldots i_s\, i)$. Now the product is

$$\lambda\, \nu(i_1 \ldots i_s)\, t_{i_1} \ldots t_{i_s} t_{j_1} \ldots {}_{j_k}$$

and clearly

$$\nu(i_1 \dots i_s) \equiv 0 \mod \Delta(i_1 \dots i_s\, j_1 \dots j_k\, i).$$

Similarly for left-hand products.

We will use the following fact about the ideal D_i namely: *If* $\ell_i^{(n)}$ *and* $\lambda_i^{(n)}$ *are two words in* $a_1 \dots a_\mu$ *with* $\ell_i^{(n)} \equiv \lambda_i^{(n)} \mod D_i$ *then* $\ell_i^{(n)}$ *and* $\lambda_i^{(n)}$ *determine the same residue classes* $\bar{\mu}(i_1 \dots i_s\, i)$ $(s < n)$.

The proofs of 1, 2, 3 and 4 will be based on the following assertions.

5. *Let* $1 + L_i$ *denote the Magnus expansion of* $\ell_i^{(n)}$ $i = 1, \dots, \mu$. *Then* $L_i t_k \equiv t_k L_i \equiv 0 \mod D_i$ *for any* t_k.

This follows from the congruences

$$\mu(i_1 \dots i_s\, i) \equiv 0 \mod \Delta(i_1 \dots i_s\, k_i)$$

and $\quad \mu(i_1 \dots i_s\, i) \equiv 0 \mod \Delta(k\, i_1 \dots i_s\, i)$.

6. *If one or more factors* t_k *are inserted in the term* $\mu(i_1 \dots i_s\, i)\, t_{i_1} \dots t_{i_s}$ *then the resulting term is congruent to zero* $\mod D_i$.

This follows from the congruences

$$\mu(i_1 \dots i_s\, i) \equiv 0 \mod \Delta(i_1 \dots i_t\, k\, i_{t+1} \dots i_s\, i).$$

7. *Let* $1 + L_j$ *denote the Magnus expansion of* $\ell_j^{(n)}$. *Then* $t_j L_j \equiv L_j t_j \equiv 0 \mod D_i$.

This follows from the congruences

$$\mu(i_1 \dots i_s\, j) \equiv 0 \mod \Delta(i_1 \dots i_s\, j\, i)$$

and $\mu(i_1 \ldots i_s j) \equiv 0 \mod \Delta(j i_1 \ldots i_s i)$.

8. $F^{(n)} \equiv 0 \mod D_i$.

This follows from the results of the previous chapters, namely $f \in F^{(n)}$ if and only if its Magnus expansion contains only terms of degree $\geq n$.

Proof of 1.

Suppose that $\ell_i^{(n)} = 1 + L_i$ is replaced by $a_j \ell_i^{(n)} a_j^{-1}$. Then L_i is replaced by

$$(1 + t_j) L_i (1 - t_j + t_j^2 - + \ldots)$$

$$= L_i + \text{terms involving } t_j L_i \text{ or } L_i t_j$$

$$\equiv L_i \mod D_i, \quad \text{by 5.}$$

Similarly for replacement by $a_j^{-1} \ell_i^{(n)} a_j$. Any conjugation is a product of the preceding examples.

Proof of 2.

Suppose that $a_j = 1 + t_j$ is replaced by $a_j' = a_k a_j a_k^{-1}$. Then

$$t_j' = (1 + t_k) t_j (1 - t_k + t_k^2 - \ldots)$$

$$= t_j + \text{terms involving } t_k t_j \text{ or } t_j t_k.$$

In the construction of the Magnus expansion of $\ell_i^{(n)}$ in terms of t_i, $i \neq j$ and t_j' it follows that the second collection of terms in the expansion of t_j' gives rise to terms which are zero $\mod D_i$, by 7. Hence the residue classes are unaltered. As in the case above, the general case is built up from a sequence of similar conjugation.

Proof of 3.

Firstly consider the identity

$$a_j \ell_j^{(n)} - \ell_j^{(n)} a_j = (1 + t_j)(1 + L_j) - (1 + L_j)(1 + t_j)$$

$$= t_j L_j - L_j t_j .$$

Then 3 follows from the congruence

$$[a_j, \ell_j^{(n)}] = 1 - (a_j \ell_j^{(n)} - \ell_j^{(n)} a_j) \ell_j^{(n)^{-1}} a_j^{-1}$$

$$= 1 - (t_j L_j - L_j t_j) \ell_j^{(n)^{-1}} a_j^{-1}$$

$$\equiv 1 \; mod \; D_i , \text{ by 7.}$$

Proof of 4.

This follows from 8 and finally proves 5.8.1 (!).

5.8.2 Examples of μ .

1. For the link given in 1 example 5.7.8 we have

$\mu(a, b) = \mu(b, a) = 2$ (linking number).

So this means that the $\overline{\mu}(x y z)$ are defined $mod\ 2$. We have

$\mu(a a a) = 1, \quad \mu(a b a) = 0, \quad \mu(b b a) = 1,$ *etc.*

2. For Milnor's link (2, 5.7.8) all $\mu = 0$.

3. For Whitehead's link

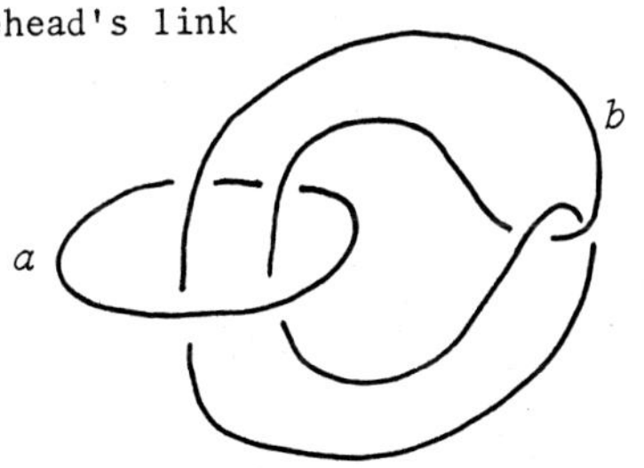

$\mu(a b b a) = \mu(b b a a) = -1 \quad \mu(b a b a) = 2$

and $\mu(a a b b) = \mu(b a a b) = -1$ and $\mu(a b a b) = 2$.

All $\mu(xy)$ and $\mu(xyz) = 0$.

4. The Sutton Hoo link [22].

$\bar{\mu}(i_1 \ldots i_k)$ $1 < k < \mu$, $\bar{\mu}(12 \ldots \mu) = 1$

$\mu(i_1 \ldots i_{\mu-2}, \mu-1, \mu) = 0$ for all other permutations of $1, \ldots, \mu-2$.

5.9 Symmetry properties of Milnor's $\bar{\mu}$ invariant.

For ij distinct it is easily seen that $\mu(ij) = \mu(k_i, k_j)$ the linking number of k_i and k_j. Also $\mu(ii) = 0$. Since linking is symmetric $\mu(ij) = \mu(ji)$. So it might be expected that the higher order $\bar{\mu}$ also contain symmetries and redundancies.

Definition 5.9.1 Proper shuffles.

By a proper shuffle $h_1 \ldots h_{r+s}$ of two sequences $i_1 \ldots i_r$ and $j_1 \ldots j_s$ will be meant one of the $(r+s)!/r!\,s!$ sequences (not all distinct) obtained by intermeshing $i_1 \ldots i_r$ with $j_1 \ldots j_s$. For example, 1213 is a proper shuffle of 11 and 23.

Theorem 5.9.2 *The following symmetry relations hold between the invariants* $\bar{\mu}$ *:*

1. *Cyclic symmetry :* $\bar{\mu}(i_1 \ldots i_r) = \bar{\mu}(i_2 \ldots i_r i_1)$.

2. *The shuffle identity :* $\sum{}' \bar{\mu}(h_1 \ldots h_{r+s} k) = 0$ *mod g.c.d.* $\Delta(h_1 \ldots h_{r+s} k)$ *where the summation extends over all proper shuffles of* $i_1 \ldots i_r$ *and* $j_1 \ldots j_s$.

Proof.

1. A Wirtinger presentation $\{a_{ij} \mid R_{ij}\}$ $i = 1, \ldots, \mu$ $j = 1, \ldots, \nu_i$ is redundant in the sense that the elements R_{ij} satisfy an identity of the form

$$\prod_{i=1}^{\mu} \prod_{j=1}^{\nu_i} \lambda_{ij} R_{ij} \lambda_{ij}^{-1} \equiv 1$$

where the λ_{ij} are words in the a_{ij}. This identity is used to attach the 3-cell to form the complementary space of the link as described in 5.4.4.

Passing to $\pi/\pi^{(n)}$ and using the presentation $P^{(n)}$ described above, we get an identity of the form

$$\prod_{i=1}^{\mu} \nu_i \left[a_i, \ell_i^{(n)} \right] \nu_i^{-1} \equiv 1 .$$

The Magnus expansion of this is

$$1 + \sum \left[t_i \, \mu(i_1 \ldots i_s i) \, t_{i_1} \ldots t_{i_s} - \mu(i_1 \ldots i_s i) \, t_{i_1} \ldots t_{i_s} t_i \right] = 1 .$$

Equating coefficients of $t_i t_{i_1} \ldots t_{i_s}$ $mod\ \Delta$ yields

$$\bar{\mu}(i_1 \ldots i_s i) = \bar{\mu}(i i_1 \ldots i_s) .$$

2. This identity follows from the shuffle identities in the Magnus expansion proved in Chen, Fox and Lyndon.

Two consequences of the above result are that $\bar{\mu}(1\,2\,3) = (-1)^{\zeta}\,\bar{\mu}(i\,j\,k)$ where $(-1)^{\zeta}$ is the sign of the permutation $\binom{1\,2\,3}{i\,j\,k}$. If $\bar{\mu}(i\,j\,k)$ is an integer (i.e. $\mu(i\,j) = \mu(j\,k) = \mu(k\,i) = 0$) then $\bar{\mu}(i\,j\,k) = 0$ if any index is repeated.

5.10 Link Homotopy

Definition 5.10.1

Two links $\ell = k_1 \cup \ldots \cup k_\mu$ and $\ell' = k_1' \cup \ldots \cup k_\mu'$ are **homotopic** if there is a continuous family of disjoint loops $\ell^{(t)} = k_1^{(t)} \cup \ldots \cup k_\mu^{(t)}$ in S^3, $0 \le t \le 1$ such that $\ell^{(0)} = \ell$ and $\ell^{(1)} = \ell'$ and such that $\ell^{(t)}$ is a link except for a finite number of exceptional values of t for which ℓ^{t-} differs from ℓ^{t+} by a simple interchange of overcrossing to undercrossing for some plane projection, see Figure 20.

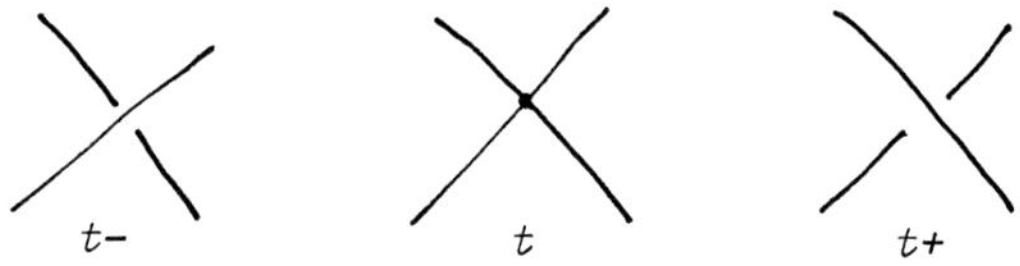

Figure 20. The basic move in link homotopy. The arcs must belong to the same component of ℓ.

This definition fits in with the philosophy that for links the tangling of individual components is uninteresting and what is important is the mutual interaction between the various components.

Example 5.10.2

1. The Whitehead link (see 5.7.5 and 5.8.2) is homotopic to the trivial link.
2. If we give the Whitehead link a little twist, then we get the link illustrated in Figure 21.

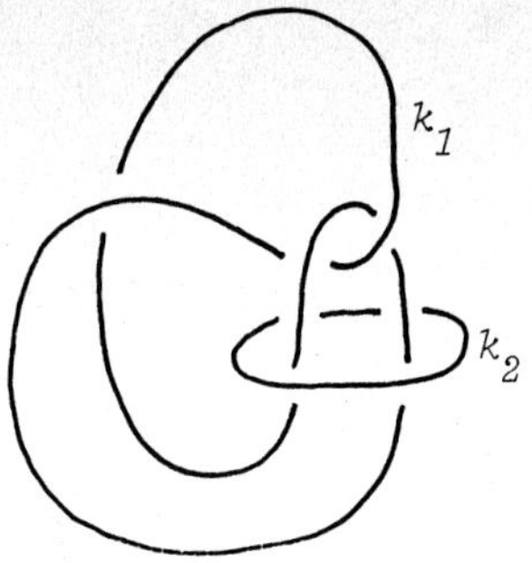

Figure 21.

This link is also homotopically trivial. However, in contrast to 1. the components k_1 and k_2 are not symetrically positioned. The component k_1 represents the trivial element in $\pi_1(S^3 - k_2)$ whilst k_2 is non-trivial in $\pi_1(S^3 - k_1)$.

The following result shows that knots are homotopically trivial in a particularly simple fashion.

Theorem 5.10.3 *Any plane projection of a knot can be made into the plane projection of the trivial knot by changing certain overcrossings into undercrossings, and vice versa.*

Proof. Orient the knot k and start tracing out the knot in a positive fashion. Whenever you come to an undercrossing proceed as follows: If the approaching overcrossing has not been visited yet, carry on, otherwise change the overcrossing to an undercrossing and carry on to the next overcrossing. After this proceedure we can describe the new knot k' as follows: Starting from an initial point the knot gradually rises and continues to rise until just before returning to the initial point, when it suddenly drops down again. It is not hard to see that k' is unknotted.

Corollary 5.10.4 *Every link is homotopic to one with unknotted components.*

Theorem 5.10.5 *The Milnor numbers $\bar{\mu}(i_1, \ldots, i_k)$ with $i_1 \ldots i_k$ distinct are invariants of link homotopy.*

Proof. Since a homotopy is a combination of link equivalence with the cutting move described in Figure 20, by 5.8.1 we need only check that μ with distinct entries is invariant under the cutting move.

There are two cases to consider.

1. *The cutting move involves the* i^{th} *component where* $i \neq i_k$.

Let $< a_i >$ denote the normal subgroup of π generated by the meridian a_i. Then under this cutting move ℓ_{i_k} is multiplied by an element x of $\left[< a_i >, < a_i > \right]$. If x is non-trivial and has Magnus expansion $1 + X$ then X contains the factor t_i at least twice and $\mu(i_1 \ldots i_k)$ is the coefficient of a term in the Magnus expansion of ℓ_{i_k} which contains the factor t_i at most once.

2. *The cutting move involves the* i_k^{th} *component.*

Then ℓ_{i_k} is multiplied by an element of the group $< a_{i_k} >$ which if non-trivial has a Magnus expansion which contains t_{i_k} at least once. On the other hand, $\mu(i_1 \ldots i_k)$ is the coefficent of a term in the Magnus expansion of ℓ_{i_k} which does not involve t_{i_k} as a factor.

5.10.6. Classifying links up to homotopy.

For links with two components the homtopy class is completely determined by the linking number. For links with three components, such that $\mu(ij) = 0$ for all ij, the homotopy class is determined by $\mu(ijk)$. For details of this and other results see Milnor, 1954. The general problem of classifying homotopy types of links seems to be a complicated problem.

Comments on Chapter 5.

1. Note the divergence here from the layman's terminology. A link should really be called a chain and each component a link. However, the use of the word chain has already been commandeered by homological algebra.

2. In Crowell & Fox and in Trotter this is shown to be a genuine weakening.

3. Conway & Mc A Gordon have associated with a knot a group, the isomorphism class of which determines the equivalence class of the knot.

4. The trivial knot or unknot 0 is not illustrated.

5. Taken from the Alexander-Briggs table.

6. See Whitehead's paper on doubled knots.

7. Indeed, if all the periods of a vector field A vanish then A must be irrotational.

8. See, for example, Flander's book on differential forms.

9. See Ferraro.

10. See Ferraro.

11. After the German mathematician who made many important discoveries in the 1930's concerning knots, 3-manifolds and topology in general. His book with Threlfall is still a classic of its kind.

12. Put $\phi_k(\infty) = 0$.

13. This result has an interesting history. The original result by Dehn in 1910 stated that a singular disc in a 3-manifold whose singularities all lay in the interior of the disc could be replaced by a non-singular disc with the same boundary. In 1929 Kneser pointed out a defect in Dehn's proof. For a long time mathematicians hunted for a correct proof, in the hope that this would turn out to be a 'philosopher's stone' turning the leaden nature of the Poincaré conjecture into pure gold. In 1957 Papakyriakopoulos gave the first correct proof. Theorem 5.2.8 is Dehn's original lemma combined with Papa's loop theorem.

There once was a lemma of Dehn
Which caused us great trial and pain
For the proof, it would topple us
Papakyriakopoulos
With a tower at last made things plain.

Fredrick Norwood.

Alternatively,

The mysterious lemma of Dehn
Drove many a good man insance
'Til Costos di Pap-
akyrikop
olous solved it without any pain.

John Milnor.

14. However, in general isomorphism of the groups does not make the knots equivalent; e.g. granny knot and reef knot.

15. The same argument applies if k is replaced by a link so that the plane projection is connected. By a preliminary isotopy we can always assume that this is the case.

16. By dual we mean under Kronecker evaluation.

17. Note that $\mathrm{curl}\, X_a = 0$ so X_a is also harmonic.

18. 5.5.12 and 5.2.9 are special cases of a more general theorem which applies in all dimensions and to all manifolds. The reader may care to note that the structure of the proof in 5.5.12 part 2 applies universally but that the existence proof, part 1, is more difficult to generalise.

19. In calculations there is no need for this, provided that the group is taken to be the free product of the groups obtained from various components.

20. See Higman 1948.

21. See Chen 1952. In Milnor's paper of 1957 these are called the link groups. Murasugi calls something slightly different the Chen groups.

22. So called because of a cauldron chain from the Sutton Hoo exhibit in the British Museum.

6. MASSEY PRODUCTS

'It has long been recognised that the homologies of differential algebras and of differential modules over differential algebras have not only products but also higher order operations, namely Massey products. These operations have largely been ignored because of the difficulty in computing them and because of their seeming lack of conceptual interest.'

J. Peter May, 1969.

A Massey triple product is defined whenever the cup product vanishes. A Massey 4-fold product is defined whenever the triple product vanishes, and so on. The Massey product is a hierarchy of increasingly discriminating algebraic constructions [1]. For example, let $[\xi]$, $[\eta]$, $[\zeta]$ be cohomology classes such that $[\xi] \cup [\eta] = 0$ and $[\eta] \cup [\zeta] = 0$. Then there are cochains x, y such that $\delta x = \xi \cup \eta$ and $\delta y = \eta \cup \zeta$.

Then with an appropriate choice of sign

$$\rho = \xi \cup y \pm x \cup \zeta$$

is a cocycle (by associativity) and so represents some cohomology class $[\rho]$.

However, this is not the end of the matter. For ρ is not necessarily uniquely defined. Let x' be any cocycle with $dim\, x' = dim\, x$ then $x + x'$ also satisfies $\delta(x+x') = \delta x = \xi \cup \eta$ and defines a possibly different product $[\rho']$.

The Massey product $<[\xi], [\eta], [\zeta]>$ is defined to be the set of all such classes $[\rho]$ which can be defined in this way.

Here are some examples.

1. Let K be the 2-complex $K = \{a, b, c \mid [[a, b], c]\}$. Let $\{u_a, u_b, u_c\}$ be the basis of $H'(K)$ dual to the 1-cells a, b, c. These classes can be represented by the geometric co-

chains ξ_a, ξ_b, ξ_c indicated in Figure 1.

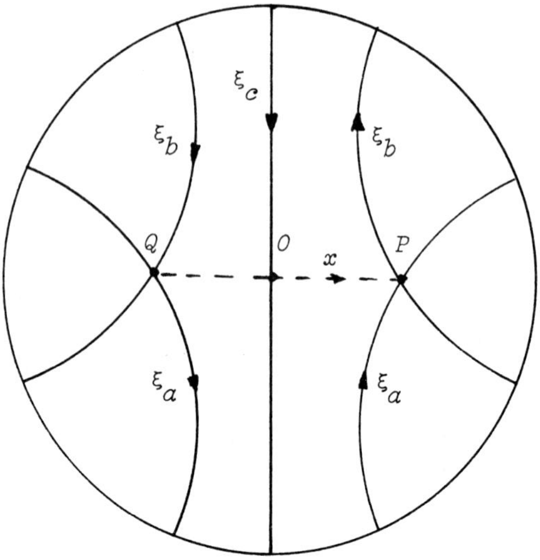

Figure 1.

The cup product $\xi_a \cup \xi_b$ is represented by the geometric cochain $P - Q$. The cup product $\xi_b \cup \xi_c$ is zero.

The cocycle $P - Q$ cobounds the geometric cochain x represented by the dotted line in the picture.

So the product $< u_a, u_b, u_c >$ contains the class of $\pm x \cup \xi_c = \pm O$ which is a generator of $H^2(K) \approx \mathbb{Z}$.

It turns out that this is the only value of the triple product in this case.

2. Suppose now that we consider the similar complex $L = \{a, b, c, d \mid [a, b] d^{-1} c d [b, a] c^{-1}\}$. Then the geometric cochain looks as in Figure 2.

Now $\xi_a \cup \xi_b = x'$ where $x' = x + \xi_d$ and $\pm x' \cup \xi_c$ represents twice a generator of $H^2(L)$.

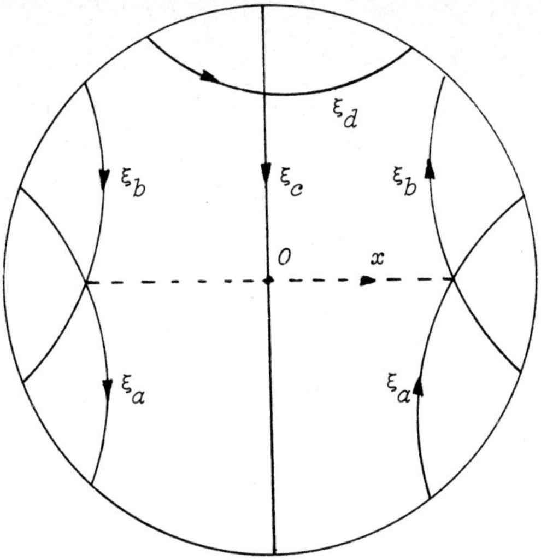

Figure 2.

By appropriate choice of x' we see that $\langle u_a, u_b, u_c \rangle = H^2(L)$.

3. Consider the Borromean rings $\ell = \ell_1 \cup \ell_2 \cup \ell_3$ spanned by three discs d_1, d_2, d_3 as in Figure 3.

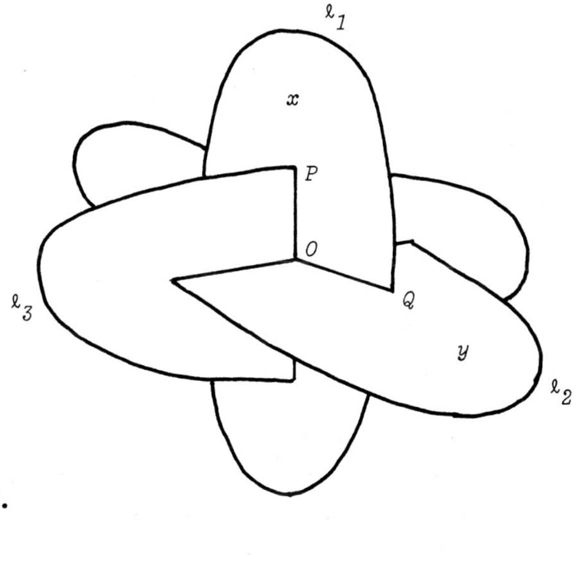

Figure 3.

By Lefschetz duality $H^1(S^3 - \ell) \approx H_2(S^3, \ell)$. The latter group has a cycle basis represented by the three discs. The dual of the cup product is represented by the transverse intersections of the discs.

Let x be the upper half of the disc d_1 so that modulo ℓ, $\partial x = d_1 \cdot d_2$. In a like manner let y be the nearer half of the disc d_2 so that $\partial y = d_2 \cdot d_3$. Then with appropriate sign convention $d_1 \cdot y + x \cdot d_3$ is the path $PO + OQ$ and this represents a basis element of $H_1(S^3, \ell)$.

The dual element in $H^2(S^3 - \ell)$ is the Massey product of the basis elements u_1, u_2, u_3 in $H^1(S^3 - \ell)$.

4. Suppose we introduce a new component ℓ_4 into the Borromean rings, as in Figure 4.

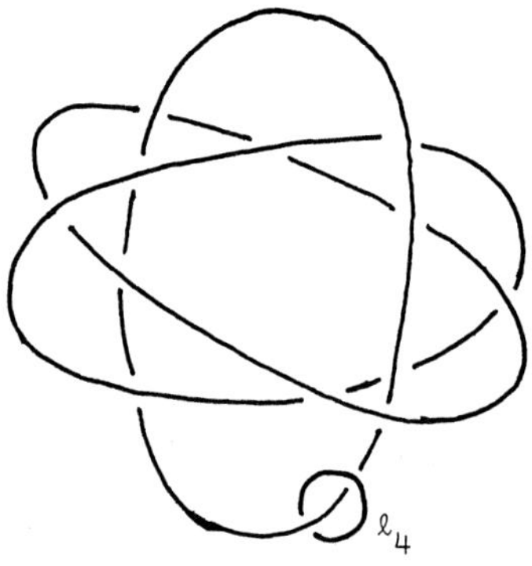

Figure 4.

Once again the Massey product $< u_1, u_2, u_3 >$ is defined, but this time it is an infinite subset of $H^2(S^3 - \ell)$.

In the sequel we shall firstly consider Massey products in 2-complexes and then relate them to the Milnor numbers of links via the theorem of Porter and Turaev.

Firstly we give a formal definition.

Definition 6.1.1 Massey products.

We will only give the definition of the Massey product $H^1 \otimes \ldots \otimes H^1 \longrightarrow H^2$ since these are the only dimensions we are interested in, [2]. Let K be a simplicial complex or regular cell complex and let $u_1, u_2, \ldots, u_n$ be elements of $H^1(K)$. A collection $a = (a_{ij})$ of cochains in $C^1(K)$, $1 \le i \le j \le n$, $i,j \ne 1,k$ is called a **defining set** for the Massey product $<u_1, \ldots, u_n>$ if the cochains satisfy the following:

1. a_{ii} is a cocycle representative of u_i, $i = 1, \ldots, u$.
2. If $\tilde{a}_{ij} = \sum_{\ell=1}^{j-1} a_{i\ell} u_{\ell+1 j}$, $1 \le i < j \le n$,

   then $\tilde{a}_{ij} = \delta a_{ij}$ $i,j \ne 1,n$.

Picture the cochains a_{ij} as lying in the upper triangular matrix

$$a = \begin{pmatrix} a_{11} & \cdots & a_{1n-1} & * \\ & \ddots & a_{ij} & \vdots \\ & & \ddots & \vdots \\ & & & a_{nn} \end{pmatrix} \quad \ldots \ldots \quad (*).$$

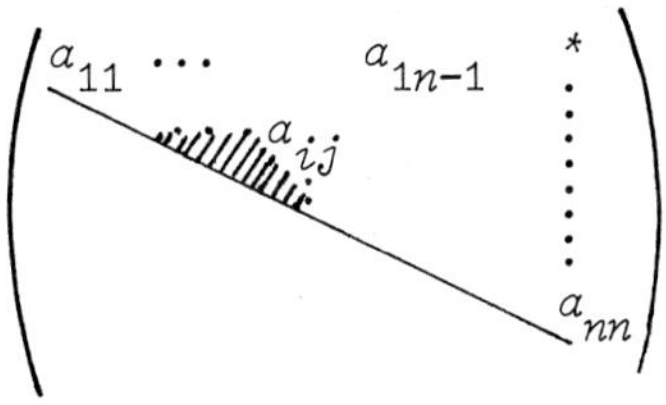

The cochain $\tilde{a}_{ij}$ is the inner product of the row to the left of a_{ij} with the column beneath a_{ij}. Then $\tilde{a}_{ij}$ $i,j \ne 1,n$ is required to be a bounding cocycle in $H^2(K)$, and we choose a_{ij} so that $\delta a_{ij} = \tilde{a}_{ij}$.

Note that this makes sense because 2. implies that

$$\delta \tilde{a}_{ij} = \sum_{\ell=1}^{j-1} \delta a_{i\ell} a_{\ell+1 j} - \sum_{\ell=1}^{j-1} a_{i\ell} \delta a_{\ell+1 j}$$

$$= \sum_{\ell=1}^{j-1} \left(\sum_{m=1}^{\ell-1} a_{im} a_{m+1 \ell} \right) a_{\ell+1 j} - \sum_{\ell=1}^{j-1} a_{i\ell} \left(\sum_{m=1}^{j-1} a_{\ell+1 m} a_{m+1} \right)$$

$$= 0.$$

since the cup product on the cochain level in a simplicial complex is associative.

The element $\tilde{a} = \tilde{a}_{1n} = \sum_{\ell=1}^{n-1} a_{i\ell} a_{\ell+1n}$ corresponding to the top right-hand corner is also a cocycle and its cohomology class $u(a)$ in $H^2(K)$ is called the **value** of the defining set a.

The **Massey product** $< u_1, \ldots, u_n >$ is said to be **defined** if it has some defining set a. If so, the Massey product is defined to be the subset of $H^2(K)$ consisting of the values $u(a)$ of all such defining sets a. In what follows the statement $< u_1, \ldots, u_n > = X$ can be taken to mean that $< u_1, \ldots, u_n >$ is defined and equals X.

The single Massey product $< u_1 >$ is just defined to be u_1 and its defining set is any cocycle representative of u_1. The cup product $u_1 u_2$ is the 2-fold Massey product $< u_1, u_2 >$. The tripe product $< u_1, u_2, u_3 >$ is defined whenever the lesser products $< u_1, u_2 >$ and $< u_2, u_3 >$ are zero. For the 4-fold product $< u_1, u_2, u_3, u_4 >$ more exacting conditions have to be satisfied for its definition.

Let a be a defining set for the Massey product $< u_1, \ldots, u_n >$. The subset of cochains $(a_{\ell m})$, $i \leq \ell \leq m \leq j$, $\ell, m \neq i, j \neq 1, n$ shown shaded in (*) is a defining set for a sub-Massey product $< u_i, \ldots, u_j >$ whose value is necessarily zero so that the cochain a_{ij} can be defined. On the other hand, we shall see later that the condition $0 \in < u_i, \ldots, u_j >$ is not sufficient for the existence of a defining set.

6.2 Some technical properties of Massey products.

In this section we examine the nuts and bolts of the previous definitions. A certain amount of care has to be taken over the details of proofs and results cannot be proved in the generality one might wish for. Consequently, on a first reading the reader may care to note the results in this section and leave the proofs until a later date.

The first result says that any collection of cocycles may be used in a defining set.

Lemma 6.2.1 *Any element of the product* $\langle u_1, \dots, u_n \rangle$ *can be defined using any representative cocycles* a_{ii} *of the cohomology classes* u_i, $i = 1, \dots, n$.

Proof. Let $a = (a_{ij})$ be a defining set for the Massey product. Then if $a_{kk} + \delta b$ is another representative for the cohomology class u_k, $1 \le k \le n$, we shall construct another defining set $a' = (a'_{ij})$ for the same product and satisfying:

(i) $a'_{ii} = a_{ii} \quad i \ne k$,

(ii) $a'_{kk} = a_{kk} + \delta b$,

(iii) $u(a') = u(a)$,

so that a and a' give the same value in the product.

Let

$$a'_{ij} = \begin{cases} a_{ik} - a_{ik-1} b, & i < k,\ j = k \\ a_{kj} + b a_{k+1j}, & j > k,\ i = k \\ a_{ij}, & i \ne k,\ j \ne k \\ a_{kk} + \delta b, & i = k,\ j = k. \end{cases}$$

A short calculation verifies that (a'_{ij}) is a defining set for $\langle u_1, \dots, u_n \rangle$ and that the following hold:

if $1 < k < n$ then $\tilde{a}'_{1n} = \tilde{a}_{1n}$,

if $k = 1$ then $\tilde{a}'_{1n} = \tilde{a}_{1n} + \delta(b a_{2n})$ and

if $k = n$ then $\tilde{a}'_{1n} = \tilde{a}_{1n} - \delta(a_{1n-1} b)$.

so that in all cases (iii) is satisfied.

Definition 6.2.2

We say that the Massey product $\langle u_1, \dots, u_n \rangle$ is **strictly defined** if each $\langle u_i, \dots, u_j \rangle = \{0\}$, $1 \le j - i \le n - 2$.

Lemma 6.2.3 *The Massey product* $<u_1, \ldots, u_n>$ *is strictly defined if and only if each partial defining set can be extended to a complete defining system for* $<u_1, \ldots, u_n>$.

Proof. Suppose that $<u_1, \ldots, u_n>$ is strictly defined. Then any partial defining set defines sub Massey products which only contain zero and so can be extended to a defining set for $<u_1, \ldots, u_n>$.

Conversely, a partial defining set can be extended if and only if the sub Massey products are zero, which implies that $<u_1, \ldots, u_n>$ is strictly defined.

Lemma 6.2.4 (Linearity). *Let* $<u_1, \ldots, u_n>$ *be defined and let* k *be an integer* $1 \leq k \leq n$.

(i) *If* $u_k = 0$ *then* $0 \in <u_1, \ldots, u_n>$

(ii) *If* λ *is an integer then*
$\lambda <u_1, \ldots, u_n> \subset <u_1, \ldots, \lambda u_k, \ldots, u_n>$.

(iii) *If* $u_k = u'_k + u''_k$ *and* $<u_1, \ldots, u'_k, \ldots, u_n>$ *is strictly defined then*
$<u_1, \ldots, u'_k + u''_k, \ldots, u_n> \subset <u_1, \ldots, u'_k, \ldots, u_n> + <u_1, \ldots, u''_k, \ldots, u_n>$.

Proof. Note firstly that (i) is a special case of (ii).

To prove (ii) let (a_{ij}) be any defining set for $<u_1, \ldots, u_n>$ and let

$$b_{ij} = \begin{cases} a_{ij} & \text{if } i > k \text{ or } j < k \\ \lambda a_{ij} & \text{if } i \leq k \leq j. \end{cases}$$

Then (b_{ij}) is a defining set for $<u_1, \ldots, \lambda u_k, \ldots, u_n>$ with $\tilde{b} = \lambda \tilde{a}$.

To prove (iii) let (a_{ij}) be any defining set for $<u_1, \ldots, u_n>$ and choose a partial defining set (b_{ij}) for $<u_1, \ldots, u'_k, \ldots, u_n>$ such that $b_{ij} = a_{ij}$ if $i > k$ or $j < k$.

Because $<u_1, \ldots, u'_k, \ldots, u_n>$ is strictly defined (b_{ij}) can be extended. Let

$$c_{ij} = \begin{cases} b_{ij} & \text{if } i > k \text{ or } j < k \\ a_{ij} - b_{ij} & \text{if } i \le k \le j . \end{cases}$$

Then (c_{ij}) is a defining set for $< u_1, \ldots, u''_k, \ldots, u_n >$ such that $\tilde{b} + \tilde{c} = \tilde{a}$.

Lemma 6.2.5 (Naturality) *Let $f : K \longrightarrow L$ be simplicial. If $< u_1, \ldots, u_n >$ is defined then so is $< f^* u_1, \ldots, f^* u_n >$ and $f^* < u_1, \ldots, u_n > \subset < f^* u_1, \ldots, f^* u_n >$.*

Proof. Let (a_{ij}) be a defining set for $< u_1, \ldots, u_n >$. If $f^{\#} : C(L) \longrightarrow C(K)$ is the cochain map induced by f then $f^{\#} a_{ij}$ is a defining set for $< f^* u_1, \ldots, f^* u_n >$. Moreover, $f^{\#} a = f^{\#} \tilde{a}$.

Definition 6.2.6 **Indeterminancy.**

If $n > 2$ one problem is that the Massey product $< u_1, \ldots, u_n >$ is not in general a unique element but rather a set of elements. The **indeterminancy** $In < u_1, \ldots, u_n >$ is defined to be $In < u_1, \ldots, u_n > = \{ a - b \mid a, b \in < u_1, \ldots, u_n > \}$.

If $< u_1, \ldots, u_n >$ is a single element then $In < u_1, \ldots, u_n > = \{0\}$ and conversely. In that situation we say that $< u_1, \ldots, u_n >$ is **uniquely defined.**

The following lemma says that $< u_1, \ldots, u_n >$ is uniquely defined if a sufficient number of smaller Massey products vanish.

Lemma 6.2.7. *If $< u_i, \ldots, u_{k-1}, v, u_{k+2}, \ldots, u_j > = \{0\}$ for all v and all $k = i, i+1, \ldots, j-1$, $1 \le i \le j \le n$. Then $< u_1, \ldots, u_n >$ is uniquely defined.*

Proof. Note that the hypotheses of the lemma imply that all products $< u_i, \ldots, u_j >$ are strictly defined.

Let (a_{ij}) and (a'_{ij}) be two defining sets for

$\langle u_1, \ldots, u_n \rangle$. By lemma 6.2.1 we may assume that $a_{ii} = a'_{ii}$, $i = 1, \ldots, n$. So $b_i = a'_{i\,i+1} - a_{i\,i+1}$ is a cocycle and represents some element v_i in H^1.

Let $(a^1_{ij}) = (a_{ij})$. By induction on k we shall obtain defining sets (a^k_{ij}) for $\langle u_1, \ldots, u_n \rangle$ and cochains (b^k_{ij}) $i \le k \le j$, $1 \le k \le n-1$ such that

$$\text{(i)} \quad a^k_{ij} = a'_{ij} \quad \text{for } j \le k, \quad b^k_{ik} = a'_{i\,k+1} - a^k_{a\,k+1},$$

$$\text{and} \quad \text{(ii)} \quad \delta b^k_{ij} = \sum_{\ell=i+1}^{k-1} a^k_{i\ell}\, b^k_{\ell+1\,j} + \sum_{\ell=k}^{j-1} b^k_{i\ell}\, a^k_{\ell+1\,j+1}$$

$$0 < j - i < n - 2.$$

Let a^k_{ij} be given. If b^k_{ij} is defined by (i) then (ii) is satisfied for $j = k$ since the right-hand side and b^k_{ik} are cocycles. The remaining b^k_{ij}, $i \le k < j$ are obtained by induction on $j - i$ as follows. Let (d^k_{ij}) be the defining set given by

$$\text{(iii)} \quad d^k_{ij} = \begin{cases} a^k_{ij} & j < k \\ b^k_{ij} & i \le k \le j \quad i' \le i \le j \le j' \\ a^k_{i+1\,j+1} & i > k. \end{cases}$$

Then (d^k_{ij}) is a defining set for $\langle u_{i'}, \ldots, v_k, \ldots, u_{j'} \rangle$ which equals $\{0\}$ by hypothesis. But (ii) is precisely the condition for (b^k_{ij}) to be a defining set for $\langle u_i, \ldots, v_k, \ldots, u_j \rangle$, so the b^k_{ij} can be defined.

Given the b^k_{ij} let

$$\text{(iv)} \quad c^k_{ij} = \begin{cases} a^k_{ij} & \text{if } j \le k \text{ or } i > k, \\ a^k_{ij} + b^k_{ij-1} & \text{if } i \le k < j-1. \end{cases}$$

Then (c^k_{ij}) is a defining set for $\langle u_1, \ldots, u_n \rangle$ such that $\tilde{c}^k = \tilde{a}^k + \tilde{d}^k$ where d^k_{ij} is given by (iii) and $1 = i'$, $n = j'$.

Formulae (ii) and (iv) imply that $a^k_{ij} = a'_{ij}$ for $j \le k+1$ so that $c^{n-1}_{ij} = a'_{ij}$. Putting all these facts together gives

$$\tilde{a}' - \tilde{a} = \tilde{c}^{n-} - \tilde{a}^1 = \sum_{k=1}^{n-1} (\tilde{c}^k - \tilde{a}^k) = \sum_{k=1}^{n-1} \tilde{d}^k .$$

Since the right-hand is a sum of coboundaries, the theorem is proved.

6.3 Some Calculations of Massey Products [3].

Definition 6.3.1 α - **complexes.**

Recall the definition of string given in Chapter 4.

A **bracket arrangement** β is a function which associates with a string I in A a commutator $\beta(I)$ in $F(A)$.

The only bracket arrangement of weight 1 is the identity. If β_1, β_2 are bracket arrangements of weight i_1 and i_2 then $[\beta_1, \beta_2]$ is a bracket arrangement of weight $i_1 + i_2$.

An α - **complex** K of weight $w(K) = n$ is a 2 - complex $K = \{A \mid \beta(I)\}$ where β is a bracket arrangement of weight $n \ge 2$ and I is a string of length n in A.

Let ξ_a denote the 1 - cochain dual to the 1 - cell $a \in A$. Then ξ_a is a cocycle as $\beta(I) \in F^n$, $n \ge 2$. Let ξ_r denote the cochain dual to the 2 - cell $r = \beta(I)$.

The following two lemmas about α - complexes can be easily proved using the techniques of Chapter 1.

Lemma 6.3.2 *Let* $K = \{A \mid \beta(I)\}$ *be an* α - *complex. Then*

(i) $H^1(K)$ *is a free Abelian group with basis the classes* $u_a = [\xi_a]$, $a \in A$.

(ii) $H^2(K)$ *is free of rank 1 with generator* $[\xi_r]$.

Let $K = \{A \mid \beta(I)\}$, $L = \{B \mid \gamma(J)\}$ be α - complexes and let $\phi_0 : F(B) \longrightarrow F(A)$ be a homomorphism of the free groups which induces a homomorphism $\phi : L \longrightarrow K$. Let $\phi^b_a = \varepsilon_a(\phi_0(b)) =$ total exponent of $a \in A$ in the word $\phi_0(b)$, $b \in B$.

Associated with $s = \gamma(J)$ in $\dot{F}(B)$ is its image $\phi_0(s)$ in $F(A)$. Then $\phi_0(s)$ is some consequence of the word r since ϕ is a homomorphism. Let λ be the total exponent of r in $\phi_0(s)$.

Lemma 6.3.3 *With the notation above, the induced maps* $\phi^* : H^1(K) \longrightarrow H^1(L)$ *and* $\phi^* : H^2(K) \longrightarrow H^2(L)$ *are given by*

$$\text{(i)} \quad \phi^*(u_a) = \sum_{b \in B} \phi_a^b \, u_b ,$$

$$\text{(ii)} \quad \phi^*[\xi_r] = \lambda[\xi_s].$$

We are now in a position to calculate Massey products in α - complexes provided that the Massey products have length no greater than the weight of the α - complex.

If I is the string $I = a_1 \ldots a_n$ let $< u_I >$ denote the product $< u_I > \; = \; < u_{a_1}, \ldots, u_{a_n} >$

Theorem 6.3.4 *If the length of the string* J *is no greater than the length of the string* I *then the Massey product* $< u_J >$ *in* $K = \{A \mid \beta(I)\}$ *is uniquely defined and has value* $\varepsilon_J(\beta(I))[\xi_r]$. *In particular, if* $\ell(J) < \ell(I)$ *then* $< u_J > \; = \{0\}$.

The proof of 6.3.4 is long and complicated and will be divided into several stages. The first stage is to define a regular cell complex which will be utilised to construct the defining sets for the Massey products.

6.3.5 The cell complex $K(\beta(I))$.

The cell complex $K(\beta(I))$, with underlying space $\{A \mid \beta(I)\}$ will be obtained after identifying the faces of a certain disc $D(\beta)$, and then wedging with a number of loops. The disc is built up from polygons lying in the hyperbolic plane [4].

Let P be a regular 8- sided polygon in the hyperbolic

plane with internal angle equal to $\pi/4$. As we saw in Chapter 3, P and its translates under the genus 2 surface group form a regular cell decomposition M of the hyperbolic plane.

The transformations we use are the following: Picture P situated centrally, with the x axis and y axis as lines of symmetry passing through the centre of four of the boundary edges. Let τ_x be a translation along the x axis through a distance equal to the width of one polygon. So P and $\tau_x P$ share a common edge which is the extreme right-hand edge of P and the extreme left-hand edge of $\tau_x P$. Let τ_y be a similar translation up the y axis.

The other symmetries considered are rotations of H about O through an angle of $\pi/4$, denoted by ω, reflection in the x axis is denoted by r_x and reflection in the y axis is denoted by τ_y.

The disc $D(\beta)$ is defined by induction on the weight of β using the cell P and some of its translates under the action of the group G generated by the above transformations.

If $\omega(\beta) = 1$ then $D(\beta) = P$. Assume now that $\omega(\beta) > 1$ and that all $D(\beta')$ for $\omega(\beta') < \omega(\beta)$ have been defined. Let $B = [\beta_1, \beta_2]$, then $D(\beta)$ is made up of five parts:

1. the central polygon P,
2. a right-hand part $\tau_x \omega D(\beta_1)$,
3. a left-hand part $r_y \tau_x \omega D(\beta_1)$,
4. an upper part $\tau_y \omega D(\beta_2)$ and
5. a lower part $r_x \tau_y \omega D(\beta_2)$.

Some idea of $D(\beta)$ can be obtained by considering a spine of $D(\beta)$ formed by the translation of portions of the x and y axis under elements of G. Some examples are given in Figure 5.

The complexes $K(\beta(I))$ are now obtained by the following identifications on the boundary of $D(\beta)$.

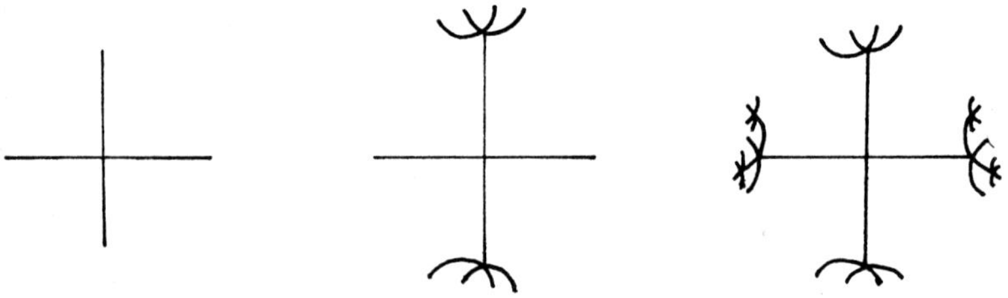

$\beta = [*, *]$ $\qquad$ $\beta = [*[*, *]]$ $\qquad$ $\beta = [[*, [*,*]], [*,*]]$

Figure 5. Spines of $D(\beta)$.

If $\beta = [a, b]$ where $a \neq b$ then identify the points of the extreme right-hand edge of $D(\beta)$ with their images under the reflection r_y. Similarly, points of the extreme top edge are identified with their images under r_x. The resulting space is a torus with a disc removed. The space $K([a, b])$ is obtained by either collapsing the hole to a point or inserting an inconsequential 2-cell and then wedging with loops corresponding to $C \in A$, $a \neq c \neq b$.

If $a = b$ then further identifications corresponding to this fact will have to be made.

If $\omega(\beta) > 2$ the identifications are carried out by induction on $\omega(\beta)$.

Let $\beta = [\beta_1, \beta_2]$ and suppose that the identifications on $D(\beta_1)$ and $D(\beta_2)$ have been defined. Then $D(\beta)$ contains a right-hand copy of $D(\beta_1)$ and its reflection in the y axis. Let p, q be two points on the right-hand copy which could be identified in $D(\beta_1)$. Then identify $p, q, r_y p, r_y q$. Proceed similarly for the upper and lower copies of $D(\beta_2)$.

If $\beta = [\beta_1, b]$ where $\omega(\beta_1) > 1$ and the identifications for $K(\beta_1)$ have been defined, we proceed in a similar fashion to the above with the top edge and bottom edge identified using r_x. Similarly if $\beta = [\,a, \beta_1\,]$.

We now consider certain cochains in M. Their algebraic properties will then determine the properties of corresponding cochains in $K(\beta(I))$.

Consider firstly the case $\beta = [\,*\,,\,*\,]$. Let ξ_x be the 1-cochain dual to the edges of $D(\beta)$ labelled ⟶ in Figure 6, and let ξ_y be the 1-cochain dual to the edges labelled ⟶⟶ .

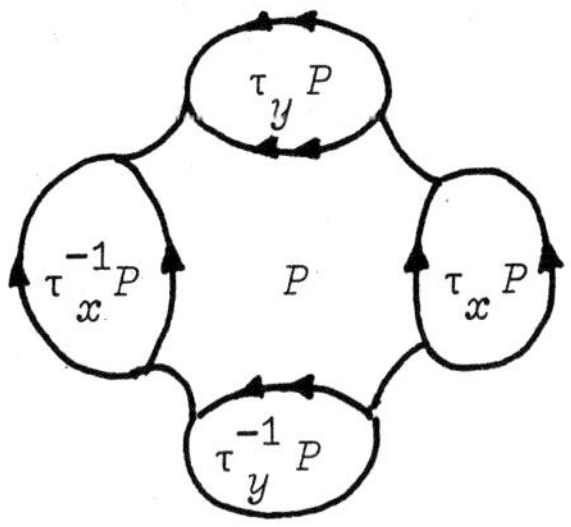

Figure 6. The cochains ξ_x and ξ_y.

Write ξ_z for ξ_x or ξ_y.. Then the coboundary of ξ_z is given by

$$\delta\,\xi_z = \xi_{P_1} - \xi_{P_2} \qquad \ldots \quad \ldots \quad \ldots \quad \ldots \qquad (1)$$

where $P_1 = \tau_z^{-2} P$ and $P_2 = \tau_z^{2} P$, $z = x$ or y.

A suitable choice of cochain approximation to the cup product is given by

$$\left.\begin{aligned} \xi_x \xi_y &= -\xi_y \xi_x = \xi_P \\ \xi_z \xi_z &= 0 \end{aligned}\right\} \qquad \ldots \quad \ldots \quad \ldots \quad \ldots \qquad (2)$$

where as usual ξ_P is the 2-cochain dual to the central cell P.

On forming $K([\,a, b\,])$, $a \neq b$, ξ_x and ξ_y correspond to certain cocycles which, by an abuse of notation, we will also denote by ξ_x and ξ_y and which have similar cup product properties.

If g is an element of the group G let $g\,\xi_z$ denote the cochain ξ_z transformed by the action of g, [5]. In the infinite cell complex M the coboundary of $g\,\xi_z$ is given by the formula

$$\delta g\,\xi_z = \xi_{gP_1} - \xi_{gP_2} \qquad \ldots \quad \ldots \quad \ldots \quad \ldots \qquad (3)$$

where $g\,P_1$ and $g\,P_2$ are the transformed polygons P_1 and P_2 of equation (1).

Equation (3) also holds in $D(\beta), (K(\beta(I))$ provided the cells $g\,P_1$ and $g\,P_2$ lie in $D(\beta), (K(\beta(I)))$.

We can represent the cochains $g\,\xi_z$ by the spines dual to their support as follows: If S_z is the portion of the z - axis ($z = x$ or y) lying in the support of ξ_z, then the spine of $g\,\xi_z$ is taken to be $g\,\xi_z$. Hence if the spine of $g\,\xi_z$ has its end points in $\partial D(\beta)$ then $g\,\xi_z$, considered as a cochain in $D(\beta)$, is a cocycle.

For a given $a \in A$ let G_a be the subset of G consisting of all those g such that the spines of $g\,\xi_z$ lie in $D(\beta)$ and their end points are identified in the cell a of $\{A \mid \beta(I)\}$. Then their sum

$$\eta_a = \sum_{g \in G_a} g\,\xi_z ,$$

taken over all distinct $g\,\xi_z$ $g \in G_a$ is a cocycle in $K(\beta(I))$ and represents the cohomology class u_a.

For example, if $\beta(I) = [a, [a, b]]$ then the spine of η_a is illustrated in Figure 7 by the heavy lines.

The reader will notice the similarity between the spine of η_a and the notion of geometric cochain discussed in Chapter 1.

Figure 7. η_a when $\beta(I) = [a,[a,b]]$.

We now consider the product structure of the $g\xi_z$. Clearly the only possible non-zero product of $g\xi_z$ with $\eta\xi_z$ will occur when the spines look like those in Figure 8.

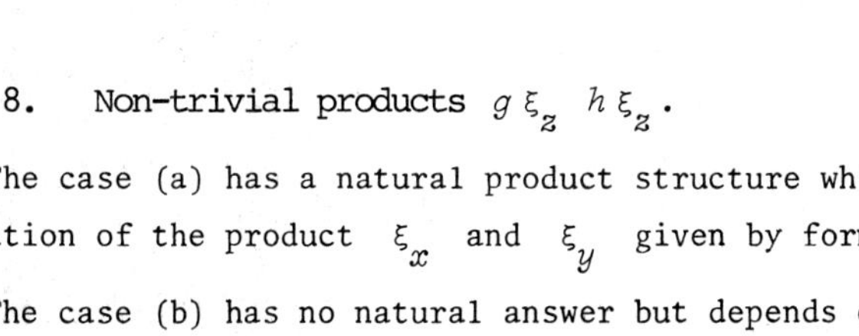

Figure 8. Non-trivial products $g\xi_z$ $h\xi_z$.

The case (a) has a natural product structure which is the translation of the product ξ_x and ξ_y given by formula (2).

The case (b) has no natural answer but depends on the ordering of the vertices of M, [6].

In order to avoid situation (b) we adopt the following device:

Definition 6.3.6 **Distinct strings.**

A string $I = a_1 a_2 \ldots a_n$ is said to be **distinct** if all the elements a_i are different, $i = 1, \ldots, n$. Similarly, $K(\beta(I))$ is said to be **distinct** if I is distinct.

Lemma 6.3.7 *If $K(\beta(I))$ is distinct and if $\ell(J) \leq \omega(\beta)$ then there is a defining set (A_{ij}) for $\langle u_J \rangle$ with value the cochain $\varepsilon_J(\beta(I))\, \xi_P$.*

Proof.

In all cases take the cocycle representatives A_{ii} for u_a to be the cocycles ξ_a defined above. We proceed by induction on $\omega(\beta)$. The induction starts easily enough when $\beta = [*, *]$. Now let $\beta = [\beta_1, \beta_2]$, $\beta(I) = [\beta_1(I_1), \beta_2(I_2)]$ where $I = I_1 I_2$. Since I is distinct the strings I_1, I_2 are distinct and so any subproduct $\langle u_{J'} \rangle$ can be defined on $K(\beta_i(I_i))$, provided $\ell(J') \leq \omega(\beta_i)$ and the value of the subproduct is $\varepsilon_{J'}(\beta_i(I_i))\, \xi_P$, $i = 1, 2$. Hence any subproduct satisfying these conditions can be defined in $K(\beta(I))$ and will be of the form $A_{ij} = a_{ij} - r_z(a_{ij})$, $z = x$ or y, where a_{ij} comes from $K(\beta_i(I_i))$, $i = 2$ or 1. If $\langle u_{J'} \rangle$ is a subproduct then the contribution from $K(\beta_i(I_i))$ will be the zero cochain if J' is not some permutation of I_i, $i = 1, 2$. So since I is distinct the value of any subproduct $\langle u_{J'} \rangle$ will have zero contribution from $K(\beta_i(I_i))$ if $\ell(J') > \omega(\beta_i)$. Therefore if $J = J_1 J_2 = J_2' J_1'$ where $\ell(J_1) = \ell(J_1') = \omega(\beta_1)$ and $\ell(J_2) = \ell(J_2') = \omega(\beta_2)$ then the only non-zero contribution to the value of (A_{ij}) can come from $\langle u_{J_1} \rangle$, $\langle u_{J_1'} \rangle$, $\langle u_{J_2} \rangle$ or $\langle u_{J_2'} \rangle$.

Since I is distinct one of J_1, J_1' is not a permutation of I_1 and one of J_2, J_2' is not a permutation of I_2. Assume initially that J_1' is not a permutation of I_1. Then $\varepsilon_{J_1'}(\beta_1(I_1)) = 0$ and the contribution of $\langle u_{J_1'} \rangle$ to (A_{ij}) will vanish.

Hence if $M = \omega(\beta_1)$ the value $\tilde{A}_{1n_1}$ is

$\varepsilon_{J_1}(\beta_1(I_1))(1-r_y)\,\xi_{P_1}$ coming from the value of the subproduct $\langle u_{J_1}\rangle$. So we may put $A_{1n_1} = \varepsilon_{J_1}(\beta_1(I_1))\,\xi_x$. Therefore the value of (A_{ij}) is

$$\begin{aligned}\tilde{A}_{1n} &= \varepsilon_{J_1}(\beta_1(I_1))\,\xi_x\ \varepsilon_{J_2}(\beta_2(I_2))\,\xi_y \\ &= \varepsilon_{J_1}(\beta_1(I_1))\,\varepsilon_{J_2}(\beta_2(I_2)) - \varepsilon_{J_1'}(\beta_1(I_1))\,\varepsilon_{J_2'}(\beta_2(I_2))\,\xi_P \\ &= \varepsilon_J(\beta(I))\,\xi_P .\end{aligned}$$

Alternatively, J_1 is not a permutation of I_1 and

$$\begin{aligned}\tilde{A}_{1n} &= -\,\varepsilon_{J_1}(\beta_1(I_1))\,\varepsilon_{J_2}(\beta_2(I_2))\,\xi_P \\ &= \varepsilon_J(\beta(I))\,\xi_P \quad \text{likewise.}\end{aligned}$$

This proves the inductive step. Notice that at no stage have we used products of type (b).

This now proves 6.3.2 in the case where I is distinct. We now relate this to the general case by the following lemma.

Lemma 6.3.8 *Given $K(\beta(I))$ there is a distinct $K(\beta(I'))$ and a map $\phi : K(\beta(I')) \longrightarrow K(\beta(I))$ which is the identity on H_2.*

Proof.

If $I = a_1 \ldots a_n$ let $I' = b_1 \ldots b_n$ be distinct. The map $K(\beta(I')) \longrightarrow K(\beta(I))$ given by $b_i \to a_i$, $i = 1, \ldots, n$ clearly satisfies the conclusion of 6.3.8.

To complete the proof of 6.3.2 assume by induction that the result is true for all J' with $\ell(J') < \ell(J)$. This means that $\langle U_J\rangle$ is both strictly and uniquely defined.

Let $K = K(\beta(I))$ and $K' = K(\beta(I'))$ and let $\phi : K' \to K$ be the map promised by 6.3.8. Then

$$< u_{a_1}, \ldots, u_{a_n} > ([K])$$

$$= < u_{a_1}, \ldots, u_{a_n} > (\phi_*[K']), \quad \text{since } \phi_*[K'] = [K]$$

$$= \phi^* < u_{a_1}, \ldots, u_{a_n} > ([K'])$$

$$= < \phi^* u_{a_1}, \ldots, \phi^* u_{a_n} > \quad \text{by naturality, 6.2.5, and the fact that}$$

$$< u_{a_1}, \ldots, u_{a_n} > \quad \text{is uniquely defined.}$$

$$= < \sum \phi_{i_1}^{j_1} u_{b_j}, \ldots, \sum \phi_{i_n}^{j_n} u_{b_{j_n}} > ([K']) \quad \text{by 6.3.3}$$

$$= \sum \phi_{i_1}^{j_1} \ldots \phi_{j_n}^{j_n} < u_{b_j}, \ldots, u_{b_{j_n}} > ([K']) \quad \text{by linearity, 6.2.4}$$

$$\sum \phi_{i_1}^{j_1} \ldots \phi_{i_n}^{j_n} \varepsilon_{j_1 \ldots j_n} (\beta(I')) \quad \text{since } I' \text{ is distinct}$$

$$\varepsilon_{i_1 \ldots i_n} (\beta(I)) \quad \text{by the naturality of the maps } \varepsilon,$$

(Chapter 4).

6.3.9 Some more examples.

1. (O'Neill) : Let $K = \left\{ a_1, \ldots, a_5 \mid [a_1, [a_2, a_3][a_1, a_5], [a_2, [a_3, a_4]][a_4, a_5] \right\}$.

Note that K has two 2-cells. Then $< u_1, u_2, u_3 >$ and $< u_2, u_3, u_4 >$ both contain zero, (due to indeterminancy). But $< u_1, u_2, u_3, u_4 >$ is not defined.

2. Let

$$K = \left\{ a_1, \ldots, a_5 \mid [a_1, [a_2, [a_3, a_4]]][a_1, [a_2, a_3]][a_1, a_5] \right\}$$

Then $< u_1, u_2, u_3, u_4 >$ is defined by not strictly defined.

6.3.10 Massey products modulo a prime.

We conclude this section by calculating Massey products in the 2-section of a lens space with coefficients in some finite cyclic group.

Theorem 6.3.11 *Let $K = \{a \mid a^p\}$ for a prime p and let k be an integer $1 \le k \le p$. Then the Massey product $< u_a, \ldots, u_a >$ of length k is defined over $\mathbb{Z}_p$ and has value*

$$< u_a, \ldots, u_a > ([Z]) = \begin{cases} 0 & 1 < k < p \\ 1 & k = p \end{cases} .$$

Proof. Let $C = \{1, t, \ldots, t^{p-1}\}$ be the multiplicative cyclic group of order p. Consider the following cochain complex $C^1 \xrightarrow{\delta} C^2 \longrightarrow 0$ with coefficients in $\mathbb{Z}$. The cochain group C^1 has a $\mathbb{Z}$-basis $\{\alpha, \beta, t\beta, \ldots, t^{p-1}\beta\}$ and has an action of C on C^1 defined by $t(\alpha) = \alpha$, $t(\beta) = t\beta$ etc. A $\mathbb{Z}$-basis for C^2 is given by $\{\sigma, t\sigma, \ldots, t^{p-1}\sigma\}$ and t acts by $t(\sigma) \doteq t\sigma$. The boundary homomorphism $\delta : C^1 \longrightarrow C^2$ is given by $\delta\alpha = (1 + t + \ldots + t^{p-1})\sigma$ and $t\beta = (1-t)\sigma$.

So an arbitrary element of C^1 is of the form $m\alpha + f(t)\beta$ and an arbitrary element of C^2 is of the form $g(t)\sigma$ where m is an integer and $f(t)$, $g(t)$ are truncated polynomials in t.

Define a product $C^1 \otimes C^1 \longrightarrow C^2$ by

$$(m_1\alpha + f_1\beta)(m_2\alpha + f_2\beta) = m_1 f_2 \sigma .$$

Now consider the element $\xi = \alpha + f_1 t$ where

$$f_1 = \sum_{j=0}^{p-2} (p-j-1)\, t^j .$$

Then $\delta\xi = p\sigma$ and so ξ is a cocycle $mod\ p$. The class $[\xi]$ generates H^1 with $\mathbb{Z}_p$ coefficients and the class $[\sigma]$ generates H^2 with either $\mathbb{Z}_p$ or $\mathbb{Z}$ coefficients. The product

$$\xi\xi = f_1(1)\sigma = \binom{p}{2}\sigma$$

More generally, let $f_k(t) = (-1)^k \sum_{j=0}^{p-k-1} \binom{p-j-1}{k} t^j$

$k = 1, 2, \ldots, p$. Then

$$(1-t)f_k = f_{k-1} + (-1)^k \binom{p}{k} \text{ and } f_k(1) = (-1)^{k+1}\binom{p}{k+1},$$
$$k > 1.$$

If $a_{ii} = \xi$, $a_{ij} = f_{j-i+1}\beta$ for $1 \le i < j \le k$, $i, j \ne 1, k$, then (a_{ij}) is a defining set for the Massey product $< [\xi], \ldots, [\xi] >$. The value of this defining set is $\binom{p}{k}\sigma$ reduced $mod\ p$ which is 0 if $1 < k < p$ and is 1 if $k = p$.

We now relate this abstract complex to $K = \{a \mid a^p\}$. We obtain K by identifying the edges of a regular p-gon Q_p in the obvious way. Any triangulation of K induces a triangulation of Q_p. Assume that K is triangulated by the simplicial complex K' so that the corresponding triangulation of Q_p is invariant under the rotation t of Q_p about its centre through an angle $2\pi/p$. Let $0 \longrightarrow D^0 \longrightarrow D^1 \longrightarrow D^2 \longrightarrow 0$ be the cochain complex of K' over $\mathbb{Z}$. Let α be dual to a 1-simplex of K' lying in an edge of Q_p. Then $\delta\alpha = (1 + t + \ldots + t^{p-1})\sigma$, where σ is dual to some 2-simplex of K' adjacent to the boundary of Q_p. Choose a 1-cochain β such that $\delta\beta = (1-t)\sigma$, see Figure 9.

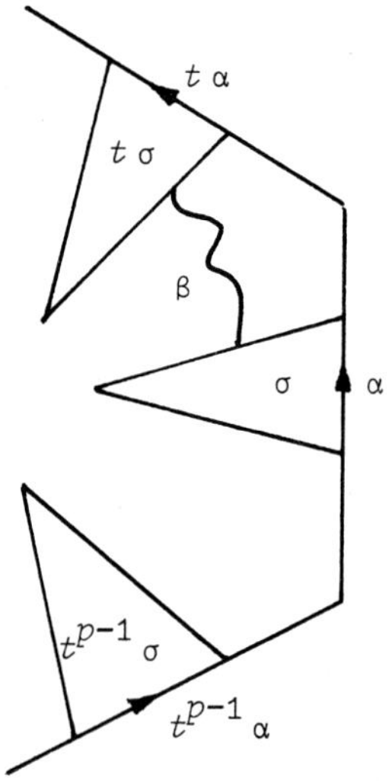

Figure 9. The cochains α, β and σ.

Let $\{D\} = C_0^1 \longrightarrow C_0^2 \longrightarrow 0$ be the cochain subcomplex of K', generated by α, β, σ and invariant under the action of t. Then the restriction $D^* \longrightarrow C_0'^*$ is a chain homotopy equivalence.

Choose an ordering of the vertices of K' which is invariant under the action of t and which induces the cup product

$$\alpha\beta = \sigma \qquad \alpha^2 = \beta^2 = \beta\alpha = 0 .$$

Then the graded algebras C_0^* and C^* are isomorphic. The calculated value of the Massey product now proves the theorem.

6.4 Massey products and the Milnor numbers.

We now use the results of the previous sections to prove a simplified version of the Porter-Turaev theorem, namely that the Milnor numbers give information about Massey products in the complement of a link.

To this end we utilise three calculating tools:

Tool 1. In this situation we have a complex which has subcomplexes which are α - complexes. e.g. let

$$K = \left\{ a_1, \ldots, a_4 \mid r_1 = [a_1, [a_2, [a_3, a_4]]], \; r_2 = [[a_1, a_2], [a_3, a_4]]^2 \right\}.$$

The subcomplexes are

$$K_1 = \left\{ a_1, \ldots, a_4 \mid r_1 \right\} \qquad K_2 = \left\{ a_1, \ldots, a_4 \mid r_2 \right\}.$$

Then the products $< u_1, \ldots, u_4 >$ can be calculated for K_1 and K_2 using 6.3.4. The product for K is clearly the union of their values on cells r_1 and r_2.

Tool 2. In this situation we have a complex whose quotients are of α - complexes. e.g. let

$$K = \left\{ a_1, \ldots, a_4 \mid [a_1, [a_2, [a_3, a_4]]]^2 \, [[a_1, a_2], [a_3, a_4]] \right\}.$$

Then $< u_1, \ldots, u_4 >$ can be calculated for

$$K_1 = \left\{ a_1, \ldots, a_4 \mid [a_1, [a_2, [a_3, a_4]]]^2 \right\} \quad \text{and}$$

$$K_2 = \left\{ a_1, \ldots, a_4 \mid [[a_1, a_2], [a_3, a_4]] \right\}$$

using 6.3.4. Their evaluations on the 2-cells are 2 and 1 respectively. So the evaluation of $< u_1, \ldots, u_4 >$ on K is $2 + 1 = 3$.

Tool 3. In this situation we ignore terms which are sufficiently far up the lower central series. We have the following lemma.

Lemma 6.4.1 *Let* $K = \{ A \mid \omega \}$ *where* $\omega \in F^{(n)}$. *If* $1 < k < n$ *then every* k*-fold Massey product vanishes on* K.

Proof. Since $\omega \in F^{(n)}$ there is a complex

$L = \{ B \mid r_1 \ldots r_i \}$ with the following properties:

1. $r_j \in F^{(n)}$, $j = 1, \ldots, i$.
2. $L_j = \{ B \mid r_j \}$ is an α-complex.
3. There is a homomorphism $f : L \to K$ of degree 1 in H_2.

Now by 1, 2 and theorem 6.3.4 all k-fold Massey products vanish on L for $1 < k < n$. Using 3. and the methods of the final part of the proof of 6.3.4 we see that all k-fold Massey products also vanish for K.

6.4.2 Massey products and links.

Let ℓ be a link. Adopting the notation developed above, let u_i be the element of H^1 determined by the ith component of ℓ and let T_i denote a hollow torus running round this component.

Theorem 6.4.3 (Porter-Turaev). *Let* ℓ *be a link with all Milnor numbers* $\mu(i_1 \ldots i_r) = 0$ *for* $1 < r < n$. *Then given any string* $j_1 \ldots j_n$, *the Massey product* $< u_{j_1}, \ldots, u_{j_n} >$ *is defined and has value* $\pm \mu(j_1 \ldots j_n)$ *on the torus* T_{j_n}.

Proof. By 6.4.1 we need only consider $S^3 - \ell$ as the space $K = \{x_i \mid [\ell_i, x_i]\}$ where $\ell_i = \ell_i^{(k)}$ and $k > n$. The generator x_i represents the ith meridian and is dual to the class u_i. The 2-cell $[\ell_i, x_i]$ corresponds to the torus T_i. Let $i = j_n$. By hypothesis and the results of Chapter 4 we see that $\ell_i \in F^{(n-2)}$ where $F = F(x_1, \ldots, x_\mu)$ so $[\ell_i, x_i] \in F^{(n-1)}$. By 6.3.4 and Tool 1 above the Massey product $\langle u_{j_1}, \ldots, u_{j_n} \rangle$ exists and is defined by the appropriate coefficient of the Magnus expansion.

Let $\ell_i = 1 + \sum \nu_{j_1 \ldots j_s} t_{j_1} \ldots t_{j_s} = 1 + L_i$ be the Magnus expansion of ℓ_i. Then $\mu(j_1 \ldots j_n) = \nu_{j_1 \ldots j_{n-1}}$. The Magnus expansion of $[\ell_i, x_i]$ is

$$1 + L_i t_i - t_i L_i$$

modulo terms of degree $\geq n+1$. This proves the theorem.

Comments on Chapter 6.

1. See Massey & Vehara.

2. The general definition is entirely analogous except for various sign conventions. For details, see Kraines or May.

3. The material for this section is taken from Fenn- Sjerve.

4. I do not know whether the hyperbolic plane has any significance other than convenience.

5. A more exact but clumsier notion is $(g^{-1})^* \xi_z$.

6. See Whitney's paper on products in 1938.

Questions for Chapter 1.

*(Questions marked with a * are very difficult. Take them as statements of proved facts, rather than exercises.)*

Cell complexes.

1. Find cell subdivisions of the projective spaces kP^n where k is R, $\mathbb{C}$, the quaternions or the Cayley numbers.

2 - complexes and group presentations.

2. Show that every compact 2- complex with only one 0- cell embeds in R^4. Give an example of one which does not embed in R^3.

3. Show that $\{a, b, c \mid abca^{-1}b^{-1}c^{-1}\}$ is the homotopy type of $S^1 \vee T^2 = \{x, y, z \mid [x, y]\}$ and hence cannot be a surface.

Homology and cohomology of cell complexes.

4. Find the homology and cohomology groups of the spaces in question 1.

5. Let Γ be the cell complex illustrated.

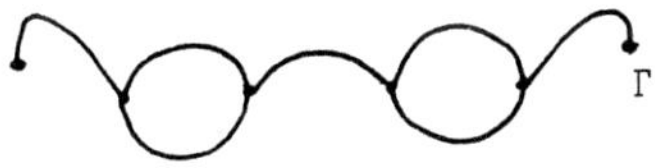

For which 1 - cells σ is ξ_σ cohomologous to zero?

(This shows that amongst cells of equal dimension in a non-manifold some are more equal than others.)

6. Show that any homology class in $H_{n-1}(M)$, where M is an n - manifold, can be represented by an embedded submanifold of dimension $n - 1$.

7.* Show that there is a 14-dimensional $\mathbb{Z}$ - homology class in a 17 - manifold which cannot be represented by an embedded submanifold.

8.* Show that there is a 9 - dimensional $\mathbb{Z}_2$ - homology class in an 11- manifold which cannot be represented by an embedded submanifold.

9.* Show that all $\mathbb{Z}_2$ - homology classes in a manifold may be represented by a continuous map from a manifold.

10.* Show that if $n \leq 6$ every n - dimensional $\mathbb{Z}$ - homology class in a manifold can be represented by a continuous map from an n - manifold.

11.* Find a counterexample to question 10 if $n > 6$.

The cup product.

12. Use the Whitney cochain formula for the cup product to show that if $\xi \in C^p(K)$, $\eta \in C^q(K)$, $\omega \in C^r(K)$ and if $f : L \longrightarrow K$ is simplicial then

1. $f^{\#}(\xi \cup \eta) = f^{\#}(\xi) \cup f^{\#}(\eta)$.

2. $\xi \cup \eta - (-1)^{pq}\, \eta \cup \xi$ is a coboundary.

3. If $1 \in C^0(K)$ is the cochain with value 1 on each vertex then $1 \cup \xi = \xi \cup 1 = \xi$.

4. $\xi \cup (\eta \cup \omega) = (\xi \cup \eta) \cup \omega$.

Existence theorems.

13. (The Borsuk-Ulam theorem). If $g : S^n \longrightarrow R^n$ is continuous then there exists a point x in S^n such that $g(x) = g(-x)$.

14.* Generalise question 13 to show that if $f : S^n \longrightarrow S^n$, $g : S^n \longrightarrow R^n$ are continuous, where the degree of f is odd, then there are points x, y in S^n such that $f(x) = -f(y)$ and $g(x) = g(y)$.

15.* Let $\{\alpha_x\}$ be a continuously varying family of paths in D^2, one for each pair $(x, -x)$ x in $\partial D^2 = S^1$, such that the endpoints of α_x are x and $-x$. Show that there are three of these paths meeting in a point.

Deduce that for any continuous map $f : M_1\, \partial M \longrightarrow D^2,\ S^1$ from the Mobius band to the disc which is a homeomorphism on the boundary there are three points m_1, m_2, m_3 in M such that $f(m_1) = f(m_2) = f(m_3)$.

16.* If $\{\beta_x\}$ is a continuously varying family of paths in D^3 one for each pair $(x, -x)$ x in $\partial D^3 = S^2$ and such that

β_x joins x and $-x$, then there are five (!) paths meeting in a common point.

17.* Let f be a continuous function which takes every line through the origin in R^3 to a parallel line.

Show that there are three lines ℓ_1, ℓ_2, ℓ_3 such that $f(\ell_1), f(\ell_2), f(\ell_3)$ are perpendicular to each other and meet in a point.

The homology and cohomology of manifolds.

18. If M is a closed compact manifold of odd dimension, show that $\chi(M) = 0$.

19. Show that $S^2 \times S^4$ and CP^3, complex projective space, have isomorphic cohomology groups but differing cohomology rings.

20. Identify generators for the cohomology and describe cup products for:

1. The 3-torus $S^1 \times S^1 \times S^1$,
2. the real projective space P^3, and
3. $P^2 \times S^1$.

21. Show that if x, y are elements of $H_p(M)$, $H_q(M)$ respectively, $p+q = \dim M$, and if u is dual to y then $x \cdot y = \pm u(x)$.

Non-orientable manifolds.

22. Triangulate the projective plane by K and represent the generator of $H_1(K)$ as a cycle in $G(K^*)$.

Explain why the dual cochain in $C^1(K)$ is not a cocycle.

Geometric cohomology.

23. Find the cohomology ring of the following complexes:

1. $\{a, b, c \mid [a, [b, c]] = 1\}$,
2. $\{a, b \mid [a, b] = 1, [a, b] = 1\}$.

24. Let X be the result of glueing a pinched torus with a hole to a Möbius band along their boundaries.

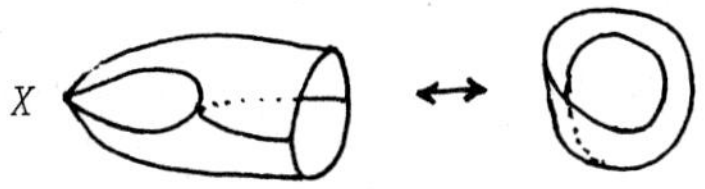

Describe the cohomology of X.

Questions for Chapter 2.

1. Give an example of a pair X, A such that

   (i) $\pi_2(X,A)$ is non-Abelian,

   (ii) $\pi_1(X,A)$ is not a group.

2. Give an example of a map $f : X \longrightarrow Y$ which induces a homology isomorphism but fails to be a homotopy equivalence.

3. Prove diagramatically that the following are presentations of the trivial group:

   (i) $\{a,b \mid a = [b, a],\ b = [a, b]\}$

   (ii) $\{a,b \mid a = b^2ab^{-3},\ b = a^2ba^{-3}\}$

   (iii) $\{a,b \mid ba\,b^{-1}a^{-1}b^3a^{-1}b^{-2}a^2ba^{-3}ba^{-3}bab^{-3}a^{-1}b^2ab^{-3}a^{-2} = 1,$
   $ab^3a^{-1}b^{-1}a^3b^{-1}a^{-2}b^2ab^{-3}ab^{-3}aba^{-3}b^{-1}a^2ba^{-3}b^{-1} = 1\}.$

4. Show that the word $aba^{-1}b^{-1}$ of length 4 is a consequence of the word $ababa$ of length 5. Represent this result diagramatically.

5. Let $K = \{a, b \mid a^2, b^3, (ab)^2\}$.

   (i) Give an example of a disc picture with boundary word $aba^{-1}b^{-2}$.

   (ii) Give an example of a spherical picture which represents a non-trivial element of $\pi_2(K)$.

   (iii) Represent the answer to (i) by a relation string and represent the answer to (ii) by an identity.

6. If $K = \{a, b, c \mid b = cac^{-1},\ c = aba^{-1},\ a = bcb^{-1}\}$, show that the Hurewicz map $h : \pi_2(K) \longrightarrow H_2(K)$ is onto. Find a spherical picture which maps under h to a generator of $H_2(K)$.

7. Show pictorially that the action on $\pi_2(K)$ is a $\pi_1(K)$ action. (i.e. show that elements of $Ker\{\pi_1(K^1) \longrightarrow \pi_1(K)\}$ act trivially.)

8. If K is a presentation of a group which has a non-trivial representation onto a subgroup of a Lie group, show that it is not possible to add a generator and relation to K so that the group becomes trivial.

9. Represent dodecahedral space as an extended group presentation.

10. Represent $\underset{k}{\#}\, S^1 \times S^2$ by an extended group presentation.

11. Show that any map $f : \underset{k}{\#}\, S^1 \times S^2 \longrightarrow \underset{k}{\#}\, S^1 \times S^2$ which induces the identity on π_1 also induces the identity on π_2 and π_3 and so is a homotopy equivalence.

12. Show that the doodle in Figure 23 is not null concordant.

13. Show that the doodle in Figure 28 is null concordant.

14. What word is read from the dotted component of the following doodle?

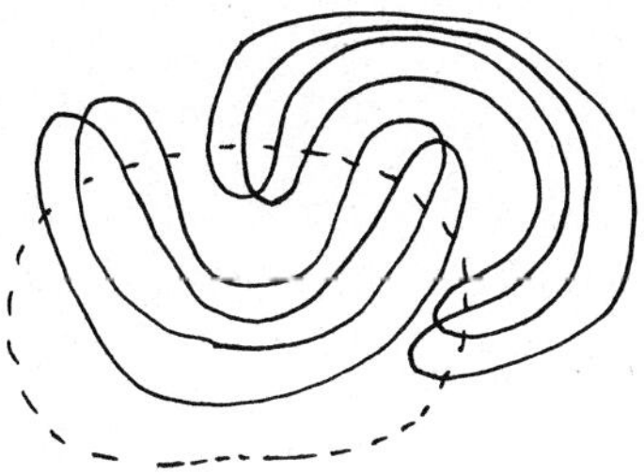

15. Show that if $i\,j\,k\,l$ are all distinct then the Borromean doodles satisfy the equation

$$(1-a_l)B_{ijk} + (1-a_i)B_{jkl} + (1-a_j)B_{kli} + (1-a_k)B_{lij} = 0$$

in Cob_n, $n \geq 4$.

16. Let $K = \{a, b \mid a^2 = b^3,\ 1 = 1\}$ and

$L = \{a, b \mid a^4b^{-3}ab^{-3}ab^{-3} = 1,\ a^4b^{-4}a^2b^{-4}a^2b^{-4} = 1\}$

Show that (i) K and L both present the same group (the trefoil group) (ii) $K \vee S^1 \simeq L \vee S^1$,

(iii) $\pi_2(K) \neq \pi_2(L)$ (so in particular $K \not\simeq L$).

Questions for Chapter 3.

1. Use the theory of covering spaces to show that there are no commutative multiplications $R^n \times R^n \longrightarrow R^n$ unless $n = 1$ or 2.

2. Calculate $\pi_2(K)$ when K is

(i) $\{a, b \mid [a, b], [a, b]\}$

(ii) $\{a \mid 1 = 1\}$.

3. Show that the fundamental group of the Hawaiian earring is not free.

4. Show that any homomorphism $\tilde{X} \longrightarrow \tilde{X}$ is an isomorphism if $\pi_1(\tilde{X}, \tilde{x}_0)$ is finite.

5. Let p_1 and p_2 be the two coverings of K illustrated.

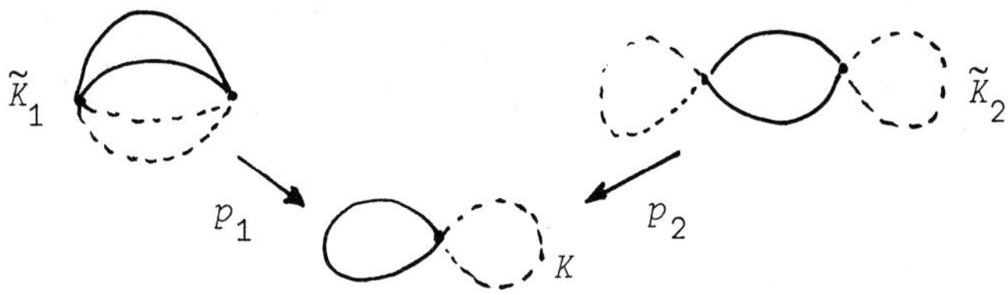

Show that there is no homomorphism $\tilde{K}_1 \longrightarrow \tilde{K}_2$.

6. Show that every homomorphism $\tilde{K} \longrightarrow \tilde{K}$ is a covering map.

7. Show that the commutator subgroup of the group $\{a, b \mid a^2 = bab^{-1}\}$ is isomorphic to the group of diadic rationals.

8. Show that the group of covering transformations of the example immediately following 3.6.9 is trivial.

9. Let $p : \tilde{X} \longrightarrow X$ be a covering. Show that p is regular if and only if $p_*\pi_1(\tilde{X}, \tilde{x}_0) = p_*\pi_1(\tilde{X}, \tilde{x}_1)$ for each lift $\tilde{x}_0, \tilde{x}_1$ where $p(\tilde{x}_0) = p(\tilde{x}_1)$.

10. Give an example of a covering $\tilde{K} \longrightarrow K$ a map $f : D \longrightarrow K$ of the disc into K and lifts $\tilde{f}_1, \tilde{f}_2$ of f such that $\tilde{f}_1$ is an embedding but $\tilde{f}_2$ is not.

11. Construct (i) the universal Abelian cover of the closed surface of genus 2; (ii) the orientation cover of the Klein bottle.

12. Show that the general form of the hyperbolic transformation having antipodal fixed points is

$$z \longrightarrow \frac{z-a}{1-\bar{a}z} .$$

Find the fixed points in this case.

13. Show that every isometry of the hyperbolic plane is of the form

$$z \longrightarrow \frac{pz+q}{\bar{q}z+\bar{p}} ,$$

where p, q are complex numbers satisfying $p\bar{p} - q\bar{q} = 1$.

14. In the isometry above, if $t = p + \bar{p}$ show that $\Delta = t^2/4$.

15. Show that if α, β are hyperbolic isometries with intersecting axes then $\alpha\beta$ is also hyperbolic.

16. Show that if α, β are hyperbolic then either $\alpha\beta$ or $\alpha\beta^{-1}$ is also hyperbolic.

17. Let α, β be isometries having
   (i) one common fixed point $\xi = \alpha\xi = \beta\xi$;
   (ii) distinct second fixed point.
Show that $\alpha\beta\alpha^{-1}\beta^{-1}$ is a non-trivial parabolic transformation with fixed point ξ.

18. Show that all parabolics are conjugate.

19. If α is hyperbolic and p parabolic with $p\xi = \xi$ then hph^{-1} is parabolic with fixed point $h\xi$.

20. Consider a pentagon $ABCDE$ in the hyperbolic plane with right angles at B, C, D and E. Show that the composition of the translation along BC through a distance $2BC$ and a translation along DE through a distance $2DC$ is a rotation about A (parabolic if A is at ∞).

21. Show that there exists a regular n-gon in H with internal angle $2\pi/m$ if and only if $m > 2n/(n-2)$.

22. Let $\theta_1,\ldots,\theta_r$ be angles $0 \le \theta_j < \pi$. Show that there exists a convex hyperbolic polygon with angles $\theta_1,\ldots,\theta_r$ if and only if

$$(r-2) > \sum_{j=1}^{m} \theta_j .$$

23. A map $f : S_1 \longrightarrow S_2$ between two closed connected surfaces is called a pinch map if there exists a subsurface X of S_1 with boundary a single loop and a point x of S_2 such that $f^{-1}x = X$ and such that $f \mid S_1 - X : S_1 - X \to X_2 - \{x\}$ is a homeomorphism. Show that any degree one map $\phi : S_1 \longrightarrow S_2$ is homotopic to a pinch map.

24. With the notation above, let $\phi : S_1 \longrightarrow S_2$ be a map with non zero degree. Show that there is a branched cover $\tilde{S}_2 \longrightarrow S_2$ and a pinch map $S_1 \longrightarrow \tilde{S}_2$ such that ϕ is homotopic to the composition

$$S_1 \longrightarrow \tilde{S}_2 \longrightarrow S_2 .$$

25. Show that the branched cover given at the very end of Chapter 3 can be described analytically by the map

$$w = \frac{1}{2} z(3 - z^2).$$

Questions for Chapter 4.

1. Describe a cell decomposition of the universal cover $\tilde{K}$ when K is

(i) $\{a, b \mid a^3 = b^4 = (ab)^2 = 1\}$;

(ii) $\{a, b, c \mid a^2 = b^3 = 1, \quad ab^2 = ba\}$;

(iii) $\{a, b, c \mid a^2 = b^2 = c^2 = (ab)^3 = (bc)^4 = (ca)^5 = 1\}$.

2. (i) Find $d_a(w)$, $d_{ab}(w)$ when $w = a^2 b a^{-1} b^{-1} a^{-1}$.

(ii) Find $\varepsilon_{abab}(r)$, $\varepsilon_{aabb}(r)$, $\varepsilon_{ababa}(r)$ when

$$r = a b a b a^{-1} b^{-1} a b^{-1} a^{-1} b a^{-1} b a b^{-1} a^{-1} b^{-1}.$$

(iii) Show that r above lies in $F^{(4)}$.

3. Prove the following identities:

(i) $[x, y^{-1}] = [y^{-1}, [y, x]]\,[x, y]^{-1}$

(ii) $[x^{-1}, y] = [x^{-1}, [y, x]]\,[x, y]^{-1}$

(iii) $[x^{-1}, y^{-1}] = [(xy)^{-1}, [x, y]]\,[x, y]$

(iv) $[xy, z] = x\,[y, z]\,x^{-1}\,[x, z]$

(v) $[z x z^{-1}, [y, z]]\,[y z y^{-1}, [x, y]]\,[x y x^{-1}, [z, x]] = 1$.

4. Show that any string $x_1 \ldots x_n$ in the set A can be cyclically permuted so that after bracketing it is a repeated string $cc \ldots c$ of basic commutators and that c is unique.

5. Consider the set of all strings in A ordered lexigraphically. Show that:

(i) $I_1 I_2 < I_1 I_3 \Leftrightarrow I_2 < I_3$.

(ii) If $I_1 < I < I_1 I_2$ then $I = I_1 I_3$ where $I_3 < I_2$.

(iii) If $I_1 < I_2$ but $\ell(I_1) > \ell(I_2)$ then $I_1 J_1 < I_2 J_2$ for all J_1, J_2.

6. Let I_c be the string corresponding to the basic commutator $c = [c_1, c_2]$. Show that the following statements are equivalent:

(i) $I = I_c$ for some c.

(ii) Either $\ell(I) = 1$ or $I = I_1 I_2$ where $I_1 = I_{c_1}$, $I_2 = I_{c_2}$ and $I_1 > I_2$.

(iii) If I' is any non-trivial cyclic permutation of I then $I' < I$.

(iv) For all factorisations $I = I_1 I_2$ there is a shuffle J of I_1, I_2 with $I > J$.

7. Show that if $I_c = I I_{c_2}$ then there exists c_1 such that $I = I_{c_1}$ and $c = [c_1, c_2]$ where c, c_1, c_2 are basic.

8. Show that if $I_1 I_2$ is greater than or equal to each of its cyclic permutations then $I_2 I_1$ is minimal among the shuffles of I_1 and I_2.

9. Show (i) $\pi^{(k)}$ is normal in π.

(ii) $\pi^{(k)}$ is normal in $\pi^{(k-1)}$.

(iii) $\pi^{(k)} / \pi^{(k+1)}$ is Abelian.

10. Show that the operations $\varepsilon_{I_c} : F^{(n)} \longrightarrow \mathbb{Z}$ where c varies over all basic commutators of weight n form a basis for $Hom(F^{(n)} / F^{(n+1)}, \mathbb{Z})$.

11. Let $h : A \longrightarrow B$ be a homomorphisms of groups which induces an isomorphism $H_1(A) \longrightarrow H_1(B)$ and a surjection $H_2(A) \longrightarrow H_2(B)$. Show that for all n h induces an isomorphism $A / A^{(n)} \longrightarrow B / B^{(n)}$.

12. Let $A^{(\infty)} = \bigcap A^{(n)}$, $B^{(\infty)} = \bigcap B^{(n)}$. Show that with the hypothesis of question 11 above h induces an injection $A / A^{(\infty)} \longrightarrow B / B^{(\infty)}$.

13. Let π be a group and let Φ_k denote the kernel of the natural map $H_2(\pi) \longrightarrow H_2(\pi / \pi^{(k-1)})$.

If f is a group homomorphism $f : G \to \pi$ which induces an isomorphism $H_1(G) \longrightarrow H_1(\pi)$, show that for $k \geq 2$ the following conditions are equivalent:

(i) f induces an epimorphism $H_2(G)/\Phi_k(G) \longrightarrow H_2(\pi)/\Phi_k(\pi)$;

(ii) f induces an isomorphism $G / G^{(k)} \longrightarrow \pi / \pi^{(k)}$.

(iii) f induces an isomorphism $H_2(G)/\Phi_k(G) \longrightarrow H_2(\pi)/\Phi_k(\pi)$ and an injection $H_2(G)/\Phi_{k+1}(G) \longrightarrow H_2(\pi)/\Phi_{k+1}(\pi)$.

14. With the notation above, show that there is a natural short exact sequence

$$1 \longrightarrow \Phi_k(\pi)/\Phi_{k+1}(\pi) \longrightarrow \Phi_k(\pi/\pi^{(k)}) \longrightarrow \pi^{(k)}/\pi^{(k+1)} \longrightarrow 1.$$

Questions for Chapter 5

1. If B_1, B_2 are harmonic vector fields, show that $B_1 \times B_2$ is harmonic if and only if the Lie bracket $[B_1, B_2] = (B_1 \cdot grad) B_2 - (B_2 \cdot grad) B_1$ vanishes.

2. Show that the vector field defined on R^3 by $B(x) = (6x_1x_2 + x_3^4, 3x_1^2 + 2x_2x_3^2, 4x_1x_3^3 + 2x_2^2x_3)$ is irrotational and find a potential ϕ such that $B = grad\,\phi$.

3. If $\phi = -1/\|x\|$ show that $grad\,\phi = x/\|x\|^3$.

4. Show that if a is any unit vector and

$$A = \frac{x \times a}{\|x \times a\|^2}\left(\frac{a \cdot x}{\|x\|} - 1\right) \qquad (x \neq \lambda a)$$

then $curl\, A = x/\|x\|^3$.

5. Show that any class in $H^1(U)$ or $H^2(U)$ can be represented by a harmonic vector field.

6. If B_k is the magnetic field at the point x due to a knot k and if r is the shortest distance from x to k show that $\lim_{r \to 0} r\|B\| = 2$.

7. Verify by direct integration the formula for the linking number when k' is the z-axis and k is the circle $x^2 + y^2 = 1$, $z = 0$.

8. Show that the Alexander polynomial of a two component link satisfies:

   (i) $\Delta(x, y) \doteq \Delta(x^{-1}, y^{-1})$;

   (ii) $\Delta(x, 1) \doteq \Delta(x) \cdot (1 - x^{\mu_{xy}})/(1 - x)$.

9. Show that the Alexander polynomial of the link given in 5.7.8 is $1 + xy$.

10. Show that for the following link

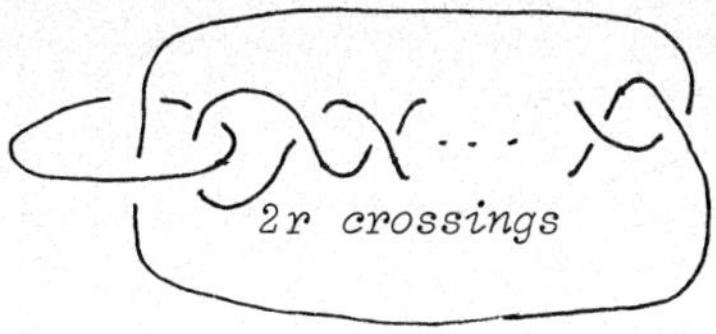

$\Delta = r(1-x)(1-y)$.

11. By calculating its Alexander module, show that Milnor's link is non-trivial.

12. In the proof of the Milnor-Chen theorem 5.7.7, show that the relations $\{R_{ij}\}$ and $\{S_{ij}\}$ are related by

$$\begin{cases} R_{i1} = S_{i1} & j = 1 \\ R_{ij} = S_{ij}(u_{ij}^{-1} S_{ij-1}^{-1} u_{ij}) & 1 < j \leq \nu_i \end{cases}.$$

13. Using the notation of 5.4.4, let ℓ_i be the length of the word e_i, $i = 1, \ldots, n$. Show that $\ell_1 + \ell_2 + \ldots + \ell_n = n$.

14. For a knot group π show that

$\pi^{(2)} = \pi^{(3)} = \pi^{(4)} = \ldots$.

15. If the orientation of the ith component of a link ℓ is reversed, show that $\bar{\mu}(i_1 \ldots i_r)$ is multiplied by $(-1)^s$ where s is the number of times the sequence $i_1 \ldots i_r$ contains i.

If the orientation of space is reversed, show that $\bar{\mu}(i \ldots i_r)$ is multiplied by $(-1)^{r-1}$.

16. If D is a doodle (Chapter 2) find a relation between $\mu(D)$ and $\mu(1\,2\,3)$ of a certain link associated with D.

17. Use 5.10.3 to unknot 10_{124}.

18. If X, Y are topological spaces and h_0, h_1 are homeomorphisms of X onto subsets of Y, show that a homotopy between h_0 and h_1 induces an isomorphism between $H^r(Y, h_0(X))$ and $H^r(Y, h_1(X))$ for all r and all homology coefficients.

Questions for Chapter 6.

1. Find all Massey triple products in the complex

$$K = \left\{ a_1, a_2, a_3 \mid a_1 a_3 a_1^{-1} a_3^{-1} a_1^{-1} a_3 a_1 a_3^{-1} a_1 a_2 a_1^{-1} a_2^{-1} a_1^{-1} a_2 a_1 a_2^{-1} \right\}.$$

2. Let $K = \{ a_1 , a_2 \mid r \}$ be a 2 - complex with $\omega(r) \in F^{(3)}$. Show that $< u_1 , u_2 , u_1 > + < u_2 , u_1 , u_1 > + < u_1 , u_1 , u_2 > = 0$.

3. Let $K = < A \mid R >$ be a 2 - complex and let $N = Ker \{ \pi(K^1) \longrightarrow \pi(K) \}$. Show that if all Massey products of length $< n$, $n > 2$, exist and are zero then $N \in F^{(n)}$ where $F = \pi(K^1)$.

4. Let $K = \{ a_1 , \ldots , a_8 \mid [a_1 , a_2][a_3 , a_4][a_5 , a_6][a_7 , a_8] \}$ be the closed orientable surface of genus 4. Show that

$$< u_1 + u_6 ,\ u_2 + u_5 ,\ u_3 + u_8 ,\ u_4 + u_7 > = H^2(K) .$$

5. If ℓ is the following link:

find all 4 - fold Massey products.

6. Let ℓ_1 , ℓ_2 be the following homotopic links:

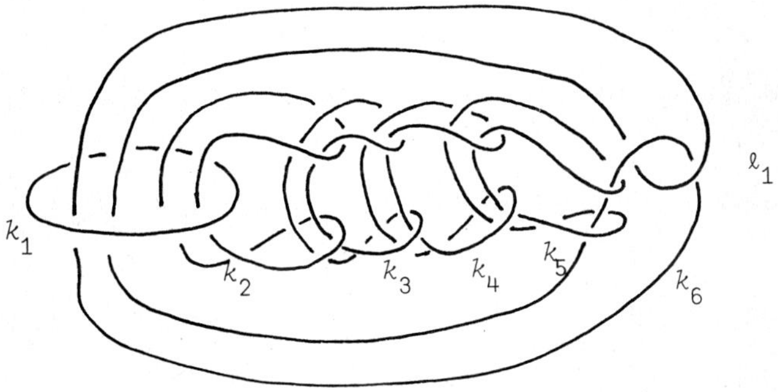

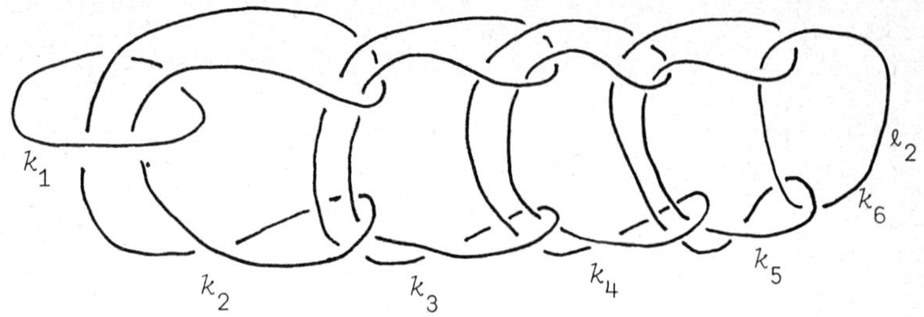

Show that for ℓ_1, $< u_1, \ldots, u_6 > \; = H^2$ and that for ℓ_2 $< u_1, \ldots, u_6 >$ only contains one element.

REFERENCES

Ahlfors L.V. and Sario L. "Riemann Surfaces" Princeton U. Press, 1960

Alexander J.W. 'Note on Riemann Spaces' Bull. Amer. Math. Soc. 26 (1919) pp 370-372.

Alexander J.W. 'On the subdivision of 3-space by a polyhedron' Proc. Nat. Acad. Sci. U.S.A. 10 (1924) pp 6-8.

Alexander J.W. 'Topological invariants of knots and links' Trans. Amer. Math. Soc. 30 (1928) pp 275-306.

Andrews J.J. and Curtis M.L. 'Free groups and handle bodies' Proc. Amer. Math. Soc. 16 (1965) pp 192-195.

Brown R. 'On the second relative homology group of an adjunction space' J. Lond. Math. Soc. (2) (1980) pp 146-152.

Brown R. and Huebschmann J. 'Identities among relations' Low dimensional topology L.M.S. Lecture Notes No.48 pp 153-202.

Buoncristiano S. Rourke C.P. and Sanderson B.J. 'A geometric approach to homology theory. L.M.S. Lecture Note series No.18.

Chen K.T. 'Commutator calculus and link invariants' Proc. Amer. Math. Soc. (1952) pp 44-55.

Chen K.T. 'Isotopy invariants of links' Ann. of Math. 56 (1952).

Chevalley C. 'Theory of lie groups" Princeton U. Press, 1946.

Cooke G.E. and Finney R.L. 'Homology of cell complexes' Princeton U. Press, 1952.

Crowell R.H. and Fox R.H. "An introduction to knot theory" Boston, Ginn 1963.

Dehn M. 'Über die Topologie des dreidimensionaler Raumes' Math. Ann. 69 (1910) pp 137-168.

Dunwoody M.J. 'The homotopy type of a 2-dimensional complex' Bull. Lond. Math. Soc. 8 (1976) pp 282-285.

Eilenberg S. and Maclane S. 'Cohomology theory in abstract groups I, II' Ann. of Math. 48 (1949) pp 51-78,326-341.

Eilenberg S. and Steenrod N. 'Foundations of Algebraic topology' Princeton U. Press, 1952.

Fenn R. and Taylor P. 'Introducing doodles' Topology of low dimensional manifolds, 1977, Ed. R. Fenn, Springer Lecture Notes 722 (1979) pp 37-43.

Ferraro V.C.A. 'Electromagnetic theory' London Athlone Press, 1954.

Fox R.H. 'On the complementary domains of a certain pair of inequivalent knots' Ned. Akad. Wetensch. Indag. Math. 14 (1952) pp 37-40.

Fox R.H. 'Free differential calculus I' Annals of Math. 57 (1953) pp 547-560.

Fox R.H. 'Covering spaces with singularities' Algebraic geometry and topology, a symposium in honor of S. Lefschetz. Princeton U. Press, 1957.

Frauz W. 'Über die Torsion einer Überdeckung' J. reine angew. Math. 173 (1935) pp 245-254.

Gauss K.F. 'Zur mathematischen Theorie der electrodynamischen Wirkungen" Werke. Königlichen Gesellschaft der Wissenschaffen zu Göttingen 5 (1877) p 605.

Hall M. 'The theory of groups' New York, Macmillan, 1959.

Hall P. 'A contribution to the theory of groups of prime power order' Proc. Lond. Math. Soc. 36 (1933) pp 29-95.

Hardy G.H. and Wright E.M. 'An introduction to the theory of numbers' Oxford, 1954.

Higman G. 'A theorem on linkages' Quart. J. Math. 19 (1948) pp 117-122.

Higman G. 'A finitely generated infinite simple group' J. London Math. Soc. 26 (1951) pp 61-64.

Hilden H.M. 'Every closed orientable 3-manifold is a 3-fold branched covering space of S^3.' Bull. Amer. Math. Soc. 80 (1974) pp 1243-1244.

Hilton P.J. and Wylie S. "Homology theory" Cambridge U. Press, 1960.

Hopf H. 'Fundamentalgruppe und zweite Bettische Gruppe' Comm. Math. Helv. 14 (1941) pp 257-309.

Kneser H. 'Geschlossene Flächen in dreidimensionalen Manningfaltigkeiten' Jehresbericht der Dent. Math. Verein. 38 (1929) 248-260.

Ky Fan 'Combinatorial properties of certain simplicial and cubical vertex maps' Arch. Math. XI (1960) pp 368-377.

Lazard M. 'Sur les groupes nilpotents et les anneaux de Lie' Annales Sci. L'Ecole Normale Superieure 71 (1954) pp 101-190.

Lefschetz S. 'Algebraic topology' Amer. Math. Soc. publications Vol. XXVII .

Little C.N. 'Alternate ± knots of order 11' Trans. Roy. Soc. Edin. 36 (1890) pp 253-255.

Lomonaco S.J. Jr. 'The second homotopy group of a spun knot' Topology 8 (1969) pp 95-98.

Magnus W., Karrass A. and Solitar D. 'Combinatorial group theory' Interscience, 1966.

Massey W.S. 'Some higher order cohomology operations' Symp. Intern. Topologia Algebraica, Mexico, 1958 UNESCO.

Massey W.S. "Algebraic topology : an introduction" Harcourt, Brace and World, Inc.

Massey W. and Vehera H. 'The Jacobi identity for Whitehead products' Algebraic geometry and topology, a symposium in honor of S. Lefschetz. Princeton U. Press, 1957.

May J.P. 'Matric Massey products' Journ. Alg. 12 (1969) pp 533-568.

Meier-Wunderli H. 'Note on a basis of P. Hall for the higher commutators in free groups' Comm. Math. Helv. 26 (1952) pp 1-5.

Milnor J. 'Link groups' Ann. of Math. 59 (1954) pp 177-195.

Milnor J. 'Isotopy of links' Algebraic geometry and topology, a symposium in honor of S. Lefschetz. Princeton U. Press, 1957.

Milnor J. 'Two complexes which are homeomorphic but combinatively distinct' Ann. of Math. 74 (1961), pp 575-590.

Milnor J. 'A duality theorem for Reidemeister torsion' Ann. of Math. 76 (1962), pp 137-147.

Moïse E.E. 'Geometric topology in dimensions 2 and 3' Springer-Verlag, 1977.

Montesinos J.M. 'A representation of closed, orientable 3-manifolds as 3-fold branched covers of S^3' Bull. Amer. Math. Soc. 80 (1974) pp 845-846.

Murasugi K. 'On Milnor's invariant for links' Trans. Amer. Math. Soc. 124 (1966) pp 94-110.

Nielsen J. 'A basis for subgroups of free groups' Math. Scand. 3 (1955) pp 31-43.

O'Neill E.J. 'On Massey products' Pacific J. Math. 76 (1978) pp 123-127.

Papakyriakopoulos C.D. 'On Dehn's lemma and the asphericity of knots' Ann. of Math. 66 (1957) pp 1-26.

Porter R. 'Milnor's $\bar{\mu}$ - invariants and Massey products' Trans. Amer. Math. Soc. 257 (1980) pp 39-71.

Reidemeister K. 'Überdeckungen von Komplexen' J. Reine Angew. Math. 173 (1935) pp 164-173.

Reidemeister K. 'Durchschnitt und Schnitt von Homotopieketten' Monabsch. Math., 48 (1939) pp 226-239.

Reidemeister K. 'Complexes and homotopy chains' Bull. Amer. Math. Soc. 56 (1950) pp 297-307.

De Rham G. 'Complexes à automorphismes et homeomorphie différentiable, Ann. Inst. Fourier, Grenoble, 2 (1950) pp 51-67.

De Rham G. 'La théorie des formes différentielles exterieures et l'homologie des variétés différentiables' Rend. Mat. e Appl. 20 (1961) pp 105-146.

Rolfsen D. 'Knots and links' Publish or Perish, 1976.

Rourke C.P. 'Presentations and the trivial group' Topology of low dimensional manifolds, 1977 Ed. R. Fenn, Springer Lecture Notes 722 (1979) pp 134-143

Seifert H. and Threlfall W. 'Lehrbuch der Topologie' Chelsea N.Y., 1947.

Stallings J.R. 'Homology and central series of groups' J. Algebra 2 (1965) pp 170-181.

Tietze H. 'Über die topologischen Invarienten mehrdimensionaler Mannigfaltigkeiten' Monatsh. Math. Phys. 19 (1908) pp 1-118.

Turaev S. 'The Milnor invariants and Massey products' Zap. Nanc. Sem. Univ. of Mat. Kogo in Stet. Acad. Nank SSSR 66 (1976) pp 189-201.

Wall C.T.C. 'Formal deformations', Proc. Lond. Math. Soc. (3)16 (1966) pp 342-352.

Whitehead J.H.C. 'Simplicial spaces, nuclei and m-groups' Proc. Lond. Math. Soc. (2) 45 (1939) pp 243-327.

Whitehead J.H.C. 'On adding relations to homotopy groups' Annals of Math. 42 (1941) pp 409-428.

Whitehead J.H.C. 'Combinatorial homotopy I, II' Bull. Amer. Math. Soc. 55 (1949) pp 213-245,453-496.

Whitney H. 'On products in a complex' Ann. of Math. 39 (1938) pp 397-432.

Whitney H. 'Geometric methods in cohomology theory' Proc. Nat. Acad. Sci. U.S.A. 33 (1947) pp 7-9.

Witt E. 'Treue Darstellung Leischen Ringe' J. Reine Agnew. Math. 177 (1937) pp 152-160.

INDEX OF NOTATION

INDEX